ISO 9000
for the
Chemical Process
Industry

Helen Gillespie

McGraw-Hill

New York San Francisco Washington, D.C. Auckland Bogotá
Caracas Lisbon London Madrid Mexioo City Milan
Montreal New Delhi San Juan Singapore
Sydney Tokyo Toronto

Library of Congress Cataloging-in-Publication Data

Gillespie, Helen Davys.
   ISO 9000 for the chemical process industry / Helen Davys
Gillespie.
      p.   cm.
   Includes bibliographical references and index.
   ISBN 0-07-024237-2 (acid-free paper)
   1. Chemical industry—Quality control—Standards. 2. ISO 9000
Series Standards. I. Title.
TP149.G486 1997
660'.068'5—dc21   97-22636
                                                   CIP

McGraw-Hill

*A Division of The **McGraw·Hill** Companies*

1 2 3 4 5 6 7 8 9 0 DOC/DOC 9 0 2 1 0 9 8 7

ISBN 0-07-024237-2

*The sponsoring editor for this book was Robert Esposito, the editing
supervisor was Patricia V. Amoroso, and the production supervisor was
Clare B. Stanley. It was set in Century Schoolbook by Victoria
Khavkina of McGraw-Hill's Professional Book Group composition unit.*

Printed and bound by R. R. Donnelley & Sons Company.

McGraw-Hill books are available at special quantity discounts to use
as premiums and sales promotions, or for use in corporate training programs. For more information, please write to the Director of Special
Sales, McGraw-Hill, 11 West 19th Street, New York, NY 10011. Or
contact your local bookstore.

Contents

Introduction

In the effort to better differentiate their products and position their companies in an increasingly competitive marketplace, more and more organizations are turning to "quality" as a means to achieve a competitive edge. But quality is no longer an intellectual concept. With the advent of ISO 9000, there is now a pragmatic means of implementing an externally audited quality system through the use of an internationally recognized standard.

Implementing a quality system that is measured against a standard has numerous ramifications. It is important to understand the underlying message of ISO 9000. It focuses on writing down your procedures and defining your policies regarding how an order is processed. It starts with taking the customer's order and follows the order through manufacturing and assembly to shipment and delivery. The underlying unwritten message is that the product should meet the customer's needs. To that end, there are several requirements, including identifying nonconforming product both in the factory and in the field. In the factory, this is covered by the requirements for identifying nonconforming product and initiating corrective action. In the field, if a customer discovers problems with the product and complains to the manufacturer, the manufacturer's ISO 9000 system requires that the customer complaint be addressed in a timely manner. Depending on the nature of the product, the problem is resolved and the customer may receive a replacement, a repair, or a new revision, generally in less than a month.

The decision to implement ISO 9000 is driven by two dominant factors: the desire to remain competitive and the desire to implement a quality system to lower operating costs. However, there are other drivers also affecting implementation. ISO 9000, or an equivalent quality system, is now a requirement to sell within Europe in certain regulated industries such as medical and toy manufacturing. It is also a de facto requirement in numerous other arenas. Many U.S. government bodies, such as the Food and Drug Administration, General Services Administration, National Aeronautics and Space Administration, Nuclear Regulatory Commission, Occupational Safety and Health Administration, Department of Defense, U.S. Postal Service, and U.S. Coast Guard are contemplating or now require ISO 9000 certification from their suppliers. In some cases, government bodies are actively developing ISO 9000 supplements to incorporate the standard into existing requirements. If ISO 9000 becomes a U.S. government purchasing requirement, then it will just be a matter of time before certification becomes a requirement to do business in the United States as well as Europe.

The surprising thing about ISO 9000 is that the standard is only nine pages long. The irritating thing is the difficulty companies have interpreting the information on those nine short pages. The reality is that every company will approach the task of interpreting and implementing ISO 9000 differently.

This book will help you implement ISO 9000. It is designed to walk you around potential roadblocks; provide timelines, checklists, and reference guides; and explain the steps involved in achieving ISO 9000 certification. It details what it takes to get started and how to interpret the elements and requirements, while setting you up with an easy-to-use policy template. But the book goes beyond the traditional scope of an introduction to and explanation of ISO 9000. Such books typically fall short of being a truly useful tool by not including implementation experiences of other organizations, and by not putting the ISO 9000 standard into the global perspective of other standards that may also apply to the organization.

You will be able to create a better, more streamlined quality management system quickly and effectively because this book provides real-world ISO 9000 implementation experiences and strategies, a discussion of the relationship of ISO 9000 to GMP, GLP, and GALP, and an overview of the issues affecting your implementation. The objective is to provide a guide to successful implementation while offering an in-depth look at the internal as well as external impact of issues surrounding implementation. After all, the more you know, the better prepared you will be.

Helen Gillespie

Acknowledgments

I would like to thank all the people who have provided information and support over the years regarding ISO 9000 and other international and national standards and regulations. You were all just great to work with. You have no idea how fired up I get after an interview and how my mind races with possibilities and story angles. I thrive on the adrenaline. The following individuals in particular have been wonderfully helpful: Phil Lofty of Morton International, whose three-hour interview on a Friday evening will never be forgotten; John Peel, Adrian Orchard, and Angela Limmer of Varian Oncology Systems (U.K.), who have been not only endlessly patient but the source of significant encouragement and information; Stephen Luke and John Moore, both formerly of Varian Sample Preparation Products, who helped me cut my teeth on what it is like to document processes to ISO 9000 (aka "the edit job from hell"); Paul Batchelder, Dennis LeTendre, Joe Fanali, and Dave Beggs of Perkin-Elmer Nelson, whose support and encouragement have helped me better understand the workings of the analytical laboratory and the importance and role of validation; David Tu, Mary Kung, and Marge Veraguth of DisCopyLabs, who insist I stay involved in the day-to-day issues of keeping ISO documentation up to date for a dynamic company (aka "the edit job from hell"); Calvin Carr of Gordon Publications, who gave me my first crack at tackling ISO 9000 in print; and Joan Moynihan of the American Chemical Society and Today's Chemist at Work, whose unflagging encouragement to scrutinize standards and regulations that affect the analytical, chemical, and medical marketplaces has led me to discover some interesting facts and meet some fascinating people.

I would like to thank some of those fascinating people, including Reginald Shaughnessy, Assistant Vice-Chair at Management of the Ontario Workplace Health and Safety Agency and Chairman of TC 176, for the interviews, advice, and insight he has provided concerning the various standards; Joe Cascio, program director for Environmental, Health and Safety Standardization at IBM and chairman of the U.S.

Technical Advisory Group (TAG) to TC 207 for taking time from his ridiculously busy schedule to discuss ISO 14000 before it was issued in draft form; Jean Artur, an environmental lawyer, teacher of environmental law at the University of Delaware, and environmental expert invited to discuss international environmental law with TC 207 for a delightful interview in the California sunshine; Stan Salot, formerly of NSAI, for the continuous information and introductions he has provided since I first took his introductory class on ISO 9000 through the University of Santa Cruz, California; and Siri Segalstad of Segalstad Consulting in Oslo, Norway, whose in-depth knowledge of the various global regulatory issues that affect the laboratory in just about any environment from analytical to quality assurance and quality control (QA/QC) to medical to clinical has been a wonderful information resource.

I would also like to thank all those patient souls who persevered through my interviews and sometimes silly questions, and who I hope to immortalize somewhat by quoting them in this book. The list includes Jim Quirk, Vice President of Technical Services and Manufacturing, Beckman Instruments; Rob Ireland, Quality and Regulatory Manager for Beckman Instruments U.K.; John Goetz, Division Manager for Bio-Rad's ECS Division in the United States; Chris Rew, Marketing Manager for Bio-Rad in the United Kingdom; Gary Miyahara, Quality Manager for Fisons Instruments in the United States; Geoff Belton, Quality Manager, Fisons U.K.; Clive Higgins, who has migrated from Fisons in the United Kingdom to Thermo Systems in the United States; and Bob Hillhouse, formerly of Fisons in the United Kingdom and now in charge of LabWare Ltd.; Will Cowan, ISO Program Manager, Hewlett Packard; Jack Kelley, Manager of Analytical Instruments, Mettler; Ron Haines, Quality Systems Manager, Thermo Separations; David Lowe, Varian Analytical Instruments' U.K. Country Manager; Marq Ransom, Director of Regulatory Affairs for Waters Chromatography; Don Hart, Assistant Director of Computer and Quality Assurance Validations at The Glaxo Inc. Research Institute (GRI); AAI's Dr. Ray Miller, Director of Information Systems, and Susan Pearce, Manager of Applications Development; and, finally, James Davies, formerly a top BSI auditor and now launched as a "quality maturity" evangelist on the world through XIS.

I would also like to thank my supporting cast of friends who believe me capable of just about anything and everything (good, that is)—may they long continue in their one-sided optimism: Ian Slade, Irwin Grater, Judith Farrell-Hawkins and Tim Hawkins, Jo Webber, Joe Golden, Gerst Gibbon, and Harmon Brown. You are all often in my thoughts and seldom far from a smile.

Helen Gillespie

What You Need to Know to Succeed

To succeed in implementing ISO 9000, more than anything else you need to understand what ISO 9000 is and what it requires. But if you are like many people, you will read the standard and still not have a clear idea of what you need to do. That is why it helps to rely on process-proven techniques and the expertise of others who have already been through an implementation.

That is the intent of this book. It was written based upon actual implementation experiences. The advice contained in the following pages has been culled not just from organizations who have implemented the standard, but from numerous in-depth interviews with Fortune 500 companies and top industry leaders who shared detailed information regarding what succeeded and why. They also provided tips and helpful information for expediting the process and ensuring that you will pass your own third-party audit.

The road to ISO success is no longer uncharted. Although the road is paved, however, you will still be susceptible to some potholes. As you start implementing ISO 9000, you'll discover a bell curve of increasing employee resistance. You will hear complaints about the amount of effort required, about the difficulty of performing additional inspections, about time spent not doing the tasks employees were hired to do, and about having to undergo internal quality audits. The peak happens when it seems that everyone in the company has declared undying hatred for ISO 9000, and even you may wonder just why you wanted to do it in the first place. But the unqualified attitude of companies who pass that peak and go on to achieve their registration certificate is that it has been worth the effort, because everyone now knows who is responsible for what. Because everyone knows what to do when something goes wrong. Because the decisions have been made and documented with the input of all departments involved. Everyone now has ownership of, or involvement with, the system and is beginning to realize just how much more efficient they can be. After a year or two, when the benefits of less rework and lower operating costs kick in, there will no longer be any question about the value of documenting processes.

To reach that point, you'll need patience and perseverance. This is not a project that can or should be turned around quickly. Eighteen months is the average implementation period. This book is designed to get you past the roadblocks and heighten your awareness of issues you will need to address along the way. This chapter covers topics involved with implementing ISO 9000, as well as a close examination of the various elements and requirements. There is advice on choosing a registrar and hiring a consultant as well as preparing for and surviving the third-party audit. This is followed by information about creating your own customized quality policy manual. There is no set format that is required, but there are certain expectations on the part of registrars. One such format is provided in Appendix B. Using a process-proven format can save you from wasting time on a format that may be difficult to administer or which inadequately covers the requirements. Finally, there is a list of registrars, information on where to go for more help, and copies of generic forms that may prove helpful.

What Is ISO 9000?

A common misconception is that ISO 9000 is a manufacturing standard or a product specification. It is neither. ISO 9000 is a set of management standards governing quality assurance. It provides a framework for a systematic approach to process management that requires an organization to keep a detailed accounting of its procedures and work. This includes documenting the processes for how a company designs, produces, inspects, packages, and installs products. The structure focuses on management of processes; the result is continuous process improvement through a quality system.

ISO 9000 was written by the International Organization for Standardization (ISO) of Geneva, Switzerland, which was founded in 1946 to develop a common set of manufacturing, trade, and communications standards that are widely used today. Standards developed by ISO are voluntary; there are no legal requirements to force a country or organization to adopt them. However, countries and industries often do adopt and attach legal requirements to ISO standards, thereby making the standards mandatory. Each country sends representatives who sit on the various committees, subcommittees, and working groups. The United States' representative to ISO is the American National Standards Institute (ANSI).

In the case of ISO 9000, ISO assigned the task of drafting the ISO 9000 series of standards to Technical Committee (TC) 176. The first ISO committee on quality, TC 176, chose the British Standard BS 5750 as their model, renaming it ISO 9000. Some 90 countries have since endorsed the standard, but generally given it a different alphanumeric designation within the particular country, often to match preexisting numerical standards criteria. In the United States, it is known as ANSI/ASQC Q9000.

While ISO 9000 was conceived as a unifying measure for the European Community (EC), it was written by an international body of nations, not all of whom intended to implement the standard immediately. When the EC adopted it so quickly, this caused consternation that Europe was establishing a "fortress" mentality toward trade. However, ISO 9000 has not locked foreign manufacturers out of Europe and is steadily gaining acceptance and implementation worldwide.

Reginald Shaughnessy, Assistant Vice-Chair at Management of the Ontario Workplace Health and Safety Agency and Chairman of TC 176, adds another point regarding the generic nature of the ISO 9000 series. "It should be borne in mind that in this system of standardization, care has been taken to avoid in detail 'how' an entity should be managed. What has been developed is a consensus of the elements of a management system describing what should be done and what should be in place in order to succeed in achieving quality in all the outcomes of work itself with all the resources that are necessary."

The five parts of the standard

While the standard is usually referred to as ISO 9000, it is, in fact, composed of five parts. ISO 9000 is essentially an introductory document which provides a statement of purpose. ISO 9001, 9002, and 9003 are three specific quality system models based on the company's function and organization. ISO 9004 is a set of guidelines for the implementation and auditing of the actual process (Fig. 1.1).

ISO 9001 is a quality system model for organizations involved in all aspects of product production, from design and development to manufacturing and installation. ISO 9002 applies to companies performing production and assembly, but not research and development or servicing. ISO 9003 offers a model for final inspection and testing only (Fig. 1.2). The specific titles for these standards are as follows:

- ISO 9001: Quality systems—Model for quality assurance in design, development, production, installation, and servicing

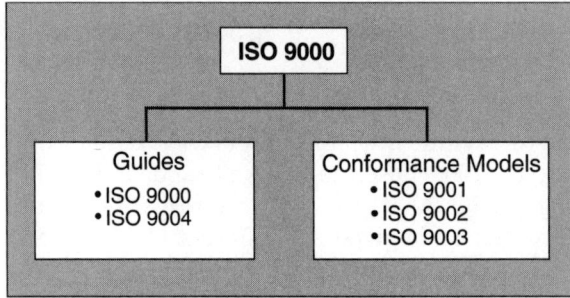

Figure 1.1 This diagram illustrates the relationships between the ISO 9000 series of standards.

Type of Standard	Name of Standard	Description of Standard
Conformance Model	ISO 9001	Quality assurance in design/development, production, installation, and servicing.
	ISO 9002	Quality assurance in production, installation, and servicing.
	ISO 9003	Quality assurance in final inspection and test.
Guide	ISO 9000	Guidelines for selection and use of the standards on quality management, quality system elements, and quality assurance.
	ISO 9004	Guidelines for quality management and quality system elements.

Figure 1.2 This table lists the three conformance models and the two guides of the ISO 9000 series of standards.

- ISO 9002: Quality systems—Model for quality assurance in production and installation

- ISO 9003: Quality systems—Model for quality assurance in final inspection and test

There are several other supporting documents and quality standards that dovetail with the ISO 9000 series and offer guidance in various aspects of quality management specific to key industries and applications, including

- ISO 8402: Quality management and quality assurance—Vocabulary

- ISO 9000-1: Quality management and quality assurance standards—Part 1: Guidelines for selection and use

- ISO 9000-3: Quality management and quality assurance standards—Part 3: Guidelines for the application of ISO 9000 to the development, supply, and maintenance of software

- ISO 9004-2: Quality management and quality system elements—Part 2: Guidelines for services

- ISO 9004-3: Quality management and quality system elements—Part 3: Guidelines for processed materials

- ISO 10011-1: Guidelines for auditing quality systems—Part 1: Auditing

- ISO 10011-2: Guidelines for auditing quality systems—Part 2: Qualification criteria for auditors

- ISO 10011-3: Guidelines for auditing quality systems—Part 3: Managing audit programs

- ISO 10012-1: Quality assurance requirements for measuring equipment—Part 1: Management of measuring equipment.

It is a good idea to review these and other models that may have an impact on your organization. Shaughnessy has stated many times in presentations around the world that "ISO 9000 must be useful. The biggest complaint is that ISO 9000 doesn't provide quality adequacy. One of the weaknesses of the series is that most people only use 9001 and 9002, and never refer to 9000 or 9004, which are the reference documents."

The differences among the system models

With the 1994 revision, all three system models are now composed of 20 elements. ISO 9001, however, covers all of the elements, while 9002 and 9003 require only those elements applicable to the specific model. The selection of 9001 versus 9002 is done not by what a company does, but by what it does not do. If the site does not perform design or research and development, it does not implement 9001 (Fig. 1.3).

One of the biggest areas of confusion surrounding the standard is the mistaken belief that 9001 is better than 9002 or 9003. It is not, nor are companies registered to 9001 "better" companies. Currently, companies that do not perform design apply for 9002. 9003 was writ-

This section...	Is covered in this ISO Standard...		
	9001	9002	9003
4.1 Management Responsibility	✓	✓	✓
4.2 Quality System	✓	✓	✓
4.3 Contract Review	✓	✓	✓
4.4 Design Control	✓		
4.5 Document and Data Control	✓	✓	✓
4.6 Purchasing	✓	✓	
4.7 Control of Customer Supplied Product	✓	✓	✓
4.8 Product Identification and Traceability	✓	✓	✓
4.9 Process Control	✓	✓	
4.10 Inspection and Testing	✓	✓	✓
4.11 Control of Inspection, Measuring, and Test Equipment	✓	✓	✓
4.12 Inspection and Test Status	✓	✓	✓
4.13 Control of Nonconforming Product	✓	✓	✓
4.14 Corrective and Preventive Action	✓	✓	✓
4.15 Handling, Storage, Packaging, Preservation, and Delivery	✓	✓	✓
4.16 Control of Quality Records	✓	✓	✓
4.17 Internal Quality Audits	✓	✓	✓
4.18 Training	✓	✓	✓
4.19 Servicing	✓	✓	
4.20 Statistical Techniques	✓	✓	✓

Figure 1.3 The 20 elements of the ISO 9000 series of standards and which element applies to the specific conformance model.

ten for companies that do not perform manufacture or assembly, just final inspection and test. It is only "easier" to achieve 9002 than 9001 because there are fewer functions involved. 9003, however, does not take into account certain quality system elements that are required by 9002 and 9001.

However, companies that perform design and servicing can apply for 9002 or 9003. Because the process of implementing ISO 9000 is difficult, some companies choose to incorporate it in stages and step up through the standard. If the company can separate these processes, this is an acceptable choice.

The 20 elements of ISO 9000

While ISO 9000 defines what a quality system should do, it does not define how to do it. That's up to the individual company. Each company does, however, need to meet certain requirements. Just seven pages long, Section Four contains the meat of the standard—the quality system requirements—which spell out the 20 areas of required conformance, labeled 4.1 through 4.20.

4.1   Management Responsibility—Requires that management define, implement, communicate, and maintain quality objectives, and assign personnel at all levels of the organization to be responsible for verifying the company's quality system. Periodic management reviews are required.

4.2   Quality System—Requires the creation and implementation of a quality manual and a quality plan, including documented procedures and instructions.

4.3   Contract Review—Requires the company to document customer orders, and to verify that it can meet customer requirements.

4.4   Design Control—Requires the documentation of quality measures in design, including design planning, input, output, review, verification, validation, and changes.

4.5   Document and Data Control—Requires procedures for creating, distributing, and tracking all documents, including changes, related to ISO 9000 activities to ensure the use of only the most current revision.

4.6   Purchasing—Establishes procedures for supplier assessment, selection, review, and monitoring, as well as verification of purchased product and component quality.

4.7   Control of Customer-Supplied Product—Stipulates how a company should handle, store, and maintain customer-supplied materials.

4.8   Product Identification and Traceability—Specifies how the company should identify products through all stages of production, delivery, and installation.

4.9   Process Control—Formalizes and ensures controlled conditions for production and installation processes that affect quality. Requires document controls and maintenance.

4.10   Inspection and Testing—Establishes procedures for inspection and testing of incoming, in-process, and outgoing products.

4.11   Control of Inspection, Measuring, and Test Equipment—Defines the requirements for equipment maintenance and calibration.

4.12   Inspection and Test Status—Establishes a system for identifying and tracking the status of products as they move through the facility.

4.13   Control of Nonconforming Product—Ensures that products not conforming to requirements are prevented from inadvertent use or installation.

4.14   Corrective and Preventive Action—Establishes, documents, and maintains procedures for investigating nonconforming products and initiating corrective and preventive action.

4.15 Handling, Storage, Packaging, Preservation and Delivery—Formalizes procedures for product handling, storage, packaging, preservation, and delivery.

4.16 Control of Quality Records—Establishes procedures for identifying, collecting, indexing, filing, storing, maintaining, and disposition of quality records.

4.17 Internal Quality Audits—Requires regular internal quality audits to ensure compliance with ISO standards. Requires corrective action of nonconformances.

4.18 Training—Identifies and provides training for all personnel performing activities that affect quality. Requires training records for verification.

4.19 Servicing—Where servicing is specified in a contract, requires procedures for performing and verifying service activities.

4.20 Statistical Techniques—Specifies the use of appropriate statistical techniques for verifying process capability.

The steps toward registration

The steps toward registration include selecting the appropriate model, assembling a steering committee, choosing a registrar and (possibly) consultants, documenting processes, implementing all new processes, and undergoing a third-party audit by the registrar. Once certification is achieved, the company continues to be audited by the registrar at regular intervals.

After achieving ISO registration, the benefits of process management by quality kick in. In fact, registration is just the first step on the path to continuous quality improvement. Organizations find they can significantly reduce the potential for production errors or rework. They are able to drive defect statistics down to impressive lows. Their streamlined processes enable reduced in-process testing and inspection, saving time and lowering costs. ISO 9000 makes continuous improvement a matter-of-fact process that delivers results to the bottom line. Companies that have been registered for several years will testify to the power and positive impact of ISO 9000 on their operations.

The Nine Steps to Registration

If you're wondering how seven short pages can be so difficult to understand, you're not alone. The best approach to take is a pragmatic one, in which each milestone is addressed step by step. As with any major project, don't rush the process, but take the time to define your needs and ensure completion at each stage.

Remember that ISO 9000 focuses on documentation. As a result, addressing the standard does not need to be a complicated process. As long as you document what you do, do what you document, and write down what you've done, you can be certified and registered.

Essentially there are nine steps to implementing ISO 9000.

1. Establish an ISO steering committee, assigning a project leader.
2. Review your current quality system, assessing what procedures are written and what need to be written.
3. Develop time lines and goals for each department.
4. Start documenting. Develop a standard format, revise existing procedures, create new ones. Create the top-level quality manual.
5. Select a registrar.
6. Assign and train internal auditors. Conduct at least one internal audit of your procedures.
7. Submit your manual for a desk audit to the registrar.
8. Undergo an audit by the registrar.
9. Receive your ISO 9000 certificate and registration.

Selecting a registrar can take place as soon as you decide to pursue certification; it is certainly not necessary to wait until you've started documenting your processes. You can also bring a consultant in at any time during the process, depending on what type of consultant you wish to use and what you need the consultant to do.

The most time-consuming portions of implementation are the document creation, of course, but also determining what needs to be done and who is going to do it. Most companies spend many hours discussing how the program should be conducted. Depending on your corporate culture, a fast "let's just do it" attitude can be as effective as a "let's think this through first" approach. The process can be expedited with the help of outside contractors who can assist with the writing and provide direction and advice. Most multinational organizations opt to register each site individually to the standard rather than go for one global registration. Using this technique, Hewlett-Packard has registered some sites in a mere six months. The time involved depends on the size of the organization and the scope of the registration.

To ISO Or Not to ISO?

There are two reasons why an organization pursues ISO 9000 registration: to achieve company marketing objectives or quality goals. Either you pursue registration for marketing reasons because your competitors are achieving certification or because you want to differentiate your company and products from the competition. Or, you pursue registration for quality reasons because you want to elevate the quality of your company's internal practices to offer a better external product or service. More often than not, it is a combination of both reasons that drives a company to implement ISO 9000. Companies contemplating registration to the standard should examine these issues and whether either or both impact their organization.

In Europe more and more companies, particularly government entities and certain regulated industries, are requiring supplier registration. It is rapidly becoming a de facto market requirement for companies that wish to do business with the EC. If two suppliers are trying to land the same EC contract, the supplier registered to ISO 9000 has a clear competitive advantage. From this standpoint, the value of an ISO 9000 registration may rest not in how much it will cost to get registered, but how much it will cost if you don't.

Beyond monetary reasons, however, the best reason to seek registration is the reason the standard was developed in the first place—to improve quality. It is designed as a framework to put in place consistent processes. It is not all-encompassing, nor is it meant to be. However, for companies just implementing quality within their organizations, it is a very workable first step.

Bottom-line benefits

Some companies also place lower operating costs on the list of reasons to implement ISO 9000. This can result in unrealistic expectations. While it is quite true that many companies have achieved outstanding reductions in operating costs and simultaneous increases in productivity during the implementation of ISO 9000, the vast majority of companies do not. Two to three years down the road they are asking "What went wrong?" But they are asking the wrong question. ISO 9000 cannot guarantee lower operating costs. In an indirect way, ISO 9000 can certainly contribute to lower operating costs when cumbersome processes are streamlined, when upper management is actively and visibly involved, and when employee process improvement suggestions are acted upon. None of these criteria are easy to achieve and can be deflected by an almost endless collection of variables. Indeed, the reasons for the inability to achieve dramatic reductions in cost can range from a history of corporate layoffs which devastate employee morale to the fact that your organization already uses streamlined processes. One of the complaints circulating among quality arenas today is that management was misled about the ability of ISO 9000 to reduce operating costs and that it is the fault of quality professionals. This is both true and false. True, because management often did have unrealistic expectations of what ISO 9000 could achieve in their organization.

False, because management tends to demand a cost justification for everything and quality professionals simply complied. Therefore, do not expect a reduction in operating costs because there is no guarantee that this will happen.

On the other hand, given the above argument, companies do find that the close scrutiny placed on processes typically enables them to streamline and reduce inefficient process costs such as rework. What they've spent to achieve registration is often being recouped in internal cost savings alone. In addition, because employees realize that the ultimate objective is process improvement, and no one wants to be improved out of a job, they take a more active role in helping to seek process improvements that meet both their own desire for employment as well as the company's desire for an improved bottom line. People can do amazing things when there's a return on their investment.

Many companies feel that the application of ISO 9000 is good commercial business sense. When lower defect rates result from ISO 9000, and the close examination of processes finds ways to double output from manufacturing, there's clear evidence that this quality process system works.

Wait and see

When would ISO 9000 registration not be a concern? If your customers don't require it. If none of your competitors are implementing it. If you're not in a regulated industry that may enforce it. If you already have a quality system in place. However, any of these parameters can change at any time.

Ultimately, the decision to implement ISO 9000 is a complex one that should come after careful consideration of the pros and cons. There is no right answer; there is only the executive-level decision to include or not include registration as part of the corporate strategy. Because it will have an impact not only on process quality but also on corporate culture, a decision to implement or not implement ISO 9000 is not one to be taken lightly.

Shaughnessy believes that quality is one of the three key pillars of any business. Together with technology and creativity, quality helps "unleash the potential of human and capital resources and create opportunities for change so necessary in today's business environment. In fact, one of the great opportunities that face business entrepreneurs and leaders is the ability to fully utilize the capability that rests in the human resources of our enterprises and radically improve the opportunity for the quality of thinking that exists in many of our entities."

2

Getting Started

How difficult is it to achieve registration to ISO 9000? If you're a Malcolm Baldrige National Quality Award winner, it might take a few weeks. If you're a medium-sized manufacturer whose quality system is nonexistent or has atrophied, it might take you several months and several thousand dollars to satisfy an auditor and receive a certificate.

It takes time to perform all the steps required to achieve an ISO 9000 registration. Not only do you need to define what needs to be done and assign responsibilities, but the documentation needs to be written and controlled; new processes required to meet various ISO 9000 elements need to be incorporated into day-to-day activities; and everything must undergo internal audits before the third-party registrar makes an appearance. Despite including the statement in the first chapter, when you hear that all you need to do is "say what you do, write it down, and do what you said you were doing," it's not quite true. That phrase is too simplistic and, as a result, misleading. While the statement *is* the essence of ISO 9000, implementing the standard entails a multitude of steps and requirements. One of the most common pitfalls occurs when organizations don't realize that "doing what you said you were doing" means that the procedures must be implemented, not just written down. A typical example happens when the registrar comes in, takes a look at the documented Internal Audit procedure, and discovers that no one has ever actually done one. The company just got their first major nonconformance. The point is not to rush the process but to heed the very words that many companies put in their quality policy: "Do it right the first time."

As with any goal, there are certain steps to take to achieve the objective. In essence, ISO 9000 registration (the goal) follows typical project management steps to achieve a documented quality system (the objective). How your organization manages this particular project depends upon your corporate culture, your organizational structure, and the ISO 9000 champions who help push the project along the path you define. Following are the key steps that every organization should take.

Assign a Project Leader

One of the first things the standard requires is the appointment of a management representative, or project leader, to champion the ISO process. The latest draft of the standard requires that this person have some clout and be someone at the executive level. The Technical Committee (TC 176) in charge of the standard added strength to this position because of complaints that the standard was difficult to implement without strong upper-management support. This is particularly necessary since the project leader chairs the ISO steering committee that spearheads the project.

In addition to "ensuring that a quality system is established, implemented, and maintained in accordance with this ISO standard," the project leader is responsible for "reporting on the performance of the quality system to the supplier's management for review and as a basis for improvement of the quality system" (4.1.2.3). These responsibilities entail not only keeping the ISO steering committee on track but also keeping upper management involved and informed before registration on the project status and after registration on the quality system's status and functionality.

Typically the QA/QC Manager takes on the role of ISO Project Leader. Organizations that don't have a formal quality department in place have elected to assign project leader status to managers in marketing, operations, document control, even human resources. Who becomes the project leader is not dictated by the standard. That this person has management authority is.

Assigning a project leader is not a point over which to agonize. Choose a person who has demonstrated leadership qualities and the ability to sustain the momentum of a complex project over time. It also helps to select someone who is respected by his or her coworkers. All the criteria that are weighed for any such appointment of responsibility are weighed in this instance as well. Once the project leader is assigned, that person's first duty is to establish the ISO steering committee.

Establish an ISO Steering Committee

Steering committees can be composed of a handful of top managers or a diverse collection of managers and supervisors. Assigning area coordinators to work with the steering committee can ensure constant communications with affected departments.

It's important to remember that you can't suddenly change the whole operation overnight. You have to consider the structure of your existing corporate organization and culture and go forward from there. Expect resistance to change. Such resistance is absolutely normal, so take the time to allow the various departments to assimilate the new quality requirements naturally. Doing so ensures that ownership for the new quality system is taken by the people who perform the processes, not just the ISO steering committee.

ISO teams can be composed of people from engineering, manufacturing, and management. Both weekly and monthly meetings can be held to discuss different levels of quality issues. Initially, it's a good idea to have biweekly or monthly meetings of the ISO team with representatives from each area. The area teams can then hold weekly meetings to address issues within their areas and perform the writing.

Sometimes a steering committee can be just a few key people who ensure that the managers who report to them are responsible for delegating down through their staff the tasks of documenting and improving processes. Sometimes the steering committee is supplemented by outside assistance, such as a consultant, that can help identify the procedures and work instructions that exist below the quality manual. Whatever the steering committee looks like, it should be based on the unique way your organization works and what makes most sense for you. As will be seen, an organization's structure, style, and corporate culture combine in a unique mix to dictate how the process is conducted. Following are three approaches that have been utilized successfully.

"There were three of us who composed the steering committee task force at Beckman," states Jim Quirk, Vice President of Technical Services and Manufacturing for Beckman Instruments. "We came up with a three-pronged approach to quality. First, that ISO would be the minimum acceptable standard for the company. Second, that we would pursue Baldrige criteria. And third, that we would implement continuous process improvement for a global quality system. To this end, division managers were made responsible, and they delegated down through their staff to document and improve processes."

Five things are always needed to achieve ISO registration, according to Marq Ransom, Director of Regulatory Affairs for Waters Chromatography. "First, documented procedures. Second, trained people. Third, good material. Fourth, good equipment both for manufacturing and gauging purposes. And fifth, a facility that fits the intent and purpose. We put together a team of managers to focus on those issues and reviewed the issues each month, particularly who was to do what and when. The project leader acts as the conduit to ensure that the managers have the resources to do the job."

Because the various units that comprise Fisons Instruments have a diverse range of business activities, Quality Manager Gary Miyahara divided the ISO certification process into three major areas with each section seeking a separate certificate. "A project leader and steering committee was formed for each group with the intent of getting most of the procedural or work instruction writing performed at the work level, and approvals made at the management level. One thing we didn't want to do was to disassociate the employee from the process."

It's important that ISO process ownership permeate all levels of the organization. Everyone will be audited, therefore everyone should be involved to some degree. As a result, it's a good idea to not only get outside training to develop an understanding of how to interpret the standards, but also to inform your organization about the changes ISO registration will require.

ISO Program Manager Will Cowan explains that Hewlett-Packard breaks the registration process into five phases: discovery, assessment, formalization, operations, and registration. Selecting the project leader and steering committee occurs during the discovery phase, in addition to identifying costs, determining the scope and level of your registration, and developing project time lines and milestones.

Cowan observes that the assessment phase is a critical part of the process, because you need to "study what you have and compare it to what you need to have. Focus on how to make it work; particularly, say what you do, do what you say, and possess data that you can prove. Almost always people rush into it and have someone write down everything. That's the wrong approach. To avoid this, we make the writing task a process improvement task. Ask yourself what you should be saying. Most important, you'll find that many processes are no longer useful and very often you just get rid of old processes." Hewlett-Packard writes the documents during the formalization phase, implements the system during the operations phase, and undergoes the audits during the registration phase.

As is obvious, no two companies are going to interpret the standard in exactly the same way. Even so, separate divisions within the same company typically use the same format or template, and often the same quality policy, to expedite implementation. Since the format is already familiar to registrars and has passed the audits, there's no reason not to leverage the design. That's part of the intent of this book. The format in Appendix B has worked for several companies in different industries. The content depends on what you do and how you can improve it.

Assess your Existing Quality System

Once the steering committee is formed and responsibilities are assigned, the next step is to assess your existing quality system. Many companies perform a preaudit at this time—either by their own employees or by a consultant or by their registrar—to determine what they have before they start documenting or changing their documentation.

The task can be approached in several ways. You could start, as many companies do, by examining the documentation currently in use. Then, appoint people who perform the tasks to review and compare these procedures to the standard. Once this is done, modify the procedures to meet any ISO requirements for that particular process. Finally, perform internal audits to hone the quality of the various processes individually and of the system as a whole. Remember to ask yourself if your company is following documented procedures during this process. In particular, look at your audit trail and make your internal audits similar to a registrar's audit by also looking at samples and reviewing key areas covered by the standard.

Some organizations take a more esoteric approach whereby they determine how to interpret the criteria by sitting back and thinking "What is good for our company?" and deciding what makes the most sense. Just be consistent in the way you handle it. Most importantly, keep it simple.

While you may discover that numerous procedures are already in place, the documentation can be inconsistent or rudimentary. The typical number of existing written procedures seems to fall in the 60 to 75 percent range. As a result, documenting the processes is often a matter of reformatting and revising existing work instructions. New procedures generally need to be implemented to address the management review, document control, and internal quality audits requirements. Other procedures may need to be expanded to address elements of the standard.

To start, you need to check what's there, what's missing, what's required, and then fill in the gaps. Thoroughly address the requirements of the standard and, in all likelihood, you'll find that you have a rather small number of new procedures that need to be written and implemented.

What causes many companies to stumble during implementation is the desire to make the perfect system when starting out. Instead, it's a better idea to seek to continuously improve quality as you go along. This is good advice for organizations in which the manufacturing units have documentation in place that addresses the U.S. Food and Drug Administration's GMPs (Good Manufacturing Practices). (More detail about the GMPs is provided in Chap. 9.) Auditing, however, will be new to other units in the company, particularly the contract review and purchasing areas. So build a framework, or expand on your existing framework, in a way that allows for flexibility.

Building this framework unearths one of the biggest complaints concerning the standard: that it doesn't require specific quality criteria such as zero defects. That's because ISO 9000 focuses on system quality, not product quality. Despite that focus, one of the first sentences in the standard under management responsibility states that "the supplier's management shall define and document its policy and objectives for, and commitment to, quality" (4.1.1). If zero defects, or its equivalent, isn't one of an organization's goals or objectives, then they're not meeting the intent of the ISO 9000 requirements. In fact, the effort required to avoid implementing processes that focus on improving product quality would not only require a genius, but be more trouble than it's worth. So, while product quality is not directly required, it will result once the appropriate processes have been defined, documented, and implemented.

In order to conduct an assessment, the scope of the quality system and the related quality activities must be defined. How this is done can pose some interesting issues. Because the organization has the option to exclude certain activities from their ISO registration, a company that designs product can elect to pursue ISO 9002 registration by eliminating design from the scope of their registration. In order to do so, that company must then treat the design department as a separate unit. Documents must be carefully controlled to ensure that the departments within the scope of the registration are able to meet the requirements without relying upon the design department. In other words, the design department cannot purchase or inspect materials that are required by other departments that will be ISO 9000–registered, since purchasing and inspection are required elements. However, because the ISO 9000 element that covers design is complex, as is the design process itself, many organizations opt to achieve ISO 9002 registration first, then incorporate the design function and get registered to ISO 9001 at a later date.

In another instance, one of the elements in 9001 and 9002 covers how products supplied by the customer are handled. This would include, for example, customer-supplied manuals for inclusion in software packaging processes. Companies seeking registration to 9002 who do not handle customer-supplied products simply state that this criterion does not apply. Since this is the case, why not have all companies apply for 9001 and state "not applicable" under areas like design and servicing? A couple of companies have been able to do just that, but not without a lot of effort and potential rejection of their registration by the accreditation boards even after the registrar indicated that they had passed the initial audit. Therefore, unless your organization has deep pockets and thrives on risk and rejection, this is not an advisable path to pursue. Indeed, the only reason a nondesign company would ask for the 9001 designation is because of the mistaken belief that 9001 is superior. This is simply not the case.

Thus, while external training and consultants can help clarify and expedite the process, understanding the standard's requirements in order to compare them to your existing system is difficult, as can be defining the scope of your registration. There's more than one way to write a quality manual and develop a system that meets the requirements. What one third-party registrar will swear to, another will swear against. Just like life, no one ever promised that implementing ISO 9000 would be easy. Hewlett-Packard's Cowan wryly comments that "Learning how to read this seven-page document is like going to law school to learn how to read the law."

Schedule a Realistic Time Line

Developing time lines and goals will fall into place once you have a clear idea of the extent of your existing documentation and what needs to be added to bring your quality system in line with the ISO 9000 standard.

The most important issue to keep in mind when scheduling a time line is to keep it realistic. Because it is central to managing the project, you will use the time line to set goals and deadlines that enable your organization to stick with the implementation program. Expect obstacles and delays. Use your energy to remove such impediments as they arise rather than to constantly readjust the time line to accommodate such delays.

What's a realistic time line in your organization depends upon your particular corporate culture. You can set up a "padded" schedule which you know you can achieve, say an 18-month plan, then do everything possible to exceed your goals. Or, you can be ambitious and set up an idealized 12-month schedule while actually expecting to take longer. Or, you can be absolutely truthful and write down an absolutely realistic schedule in the expectation that the various departments will address the elements on schedule. It depends on whether your organization operates better under the carrot or the whip. Given today's lean multitasking environment, however, it's not likely that projects will get turned around ahead of schedule. In reality, ISO 9000 projects are rarely given top priority. Don't despair, just be realistic and expect this to be one of the challenges under which you operate.

Define specific goals to achieve each month. For instance, you might decide to allow one month for the initial assessment. During that time expect the affected departments to produce all written documentation within the first week or two so that the steering committee can assess what needs to be done in the second half of the month. Does this sound like too little time for a committee that meets once a week, or only four times that month? Then extend the assessment deadline. Next, on a department-by-department basis, start with the (relatively) easy tasks first, by defining the documentation format and rewriting existing documentation to that layout. Then tackle new procedures once you're familiar and comfortable with the documentation process. Don't forget to assign goals that can be accomplished simultaneously as well as sequentially.

Often not rushing into documenting your processes can increase the quality of the end result. This is because allowing the various departments or units to implement the standard at their own speed promotes ownership of the process. Developing time lines in an interactive way such as this, rather than delegating specific due dates, has been the key to success for some organizations.

Another approach is to break down the tasks into small, easily manageable fragments. "You have to let people come at it slowly," explains Dave Beggs, Quality Assurance (QA) Manager for Perkin-Elmer Nelson. "Our software quality program was toughest. The QA gave the group an outline and suggested they start bit by bit to address various blocks of activity. Subsequently, each week the group met with the QA to discuss one of those blocks. They would come back with a draft, and it would get discussed a little more until the procedure manual was developed. Then based on that manual, they started hitting the specifics, such as developing a procedure for backup, a procedure for configuration control, a procedure for security, a procedure for viruses, etc. They just kept working at it till they got it done."

While continuous improvement is not explicitly spelled out by the standard, quality objectives are. If you're ambitious, you may add goals beyond the minimum requirements.

Perkin-Elmer Nelson, for instance, went beyond the 20 criteria and documented safety, their disaster contingency plan, and customer support. "We felt these areas were core to our business, particularly in regard to how our customers are considered," Beggs adds. "We also had some objectives for our Software Engineering Manual, particularly that it must focus on improving the quality of our products and the productivity of the engineering organization. We wanted to prevent defects and place greater emphasis on the front end of the development cycle."

Documenting processes can be an instrument of change. "We spent about a year and a half to ensure that all lower-level documents were completed, then we changed almost everything," recalls John Peel, Country Manager for Varian Oncology Systems (U.K.). "We used ISO to rewrite responsibilities, so we broke one of the golden rules that says just write down what you do. We did this because we viewed GMP as a dinosaur from the past due to its formal 'you will do this' approach that was done for defense issues."

"Now, we've told our suppliers to become certified," Peel continues. "We took our original 400 suppliers down to 100, and 51 of those are now certified. But the process is interactive. Free of charge, we're training the remaining suppliers on what to do. We're shooting for total supplier certification but we expect there will always be a few who, for one reason or another, don't pursue ISO certification."

Creating time lines is a balancing act. You must balance the amount of time spent on one task with the amount of time spent on another. More time on one task means less time on another. Determining which goal will require more time than another requires a combination of experience and knowledge of the various groups chartered with implementing the goal, as well as soothsayer skills mixed with a heavy dab of luck and timing. Each affects the others. As you develop time lines and goals, it's important to note that your ISO 9000 project is not a one-shot deal but the beginning of a long-term process. It will eventually define the way your organization does business. It's certainly not window dressing.

Leverage the Experience of Others

The first time you read the ISO 9000 standard, don't be surprised if you don't immediately understand what's expected of you. Most people don't. In fact, each time you read the standard you will understand it a little more, interpret it a little differently. Unfortunately, it wasn't written in a very user-friendly style. And because it was modeled on a British standard, terminology is used that is specific to the United Kingdom rather than the United States—an aspect that has tripped up more than one company during the course of implementation. For instance, the section on contract review actually pertains to purchasing, not to legal activities, since the British standard views purchase orders as contracts. Even so, the British have just as much difficulty interpreting and implementing the standard as any other nationality regardless of this slight verbal advantage.

Because the standard was written to be generic to all industries, it is not always clear and contains many ambiguities that can be difficult to interpret. Therefore, it behooves you to learn as much as possible about the standard early in the process so that you don't create processes and documentation that are difficult to audit, or find yourself arguing a losing battle with your external auditor. Where do you start? Luckily, ISO 9000 has been around for a few years now, and there are numerous pioneers who have already been the target of most of the arrows. Learn from their experience. Research what's been written about ISO 9000 to uncover key issues, hot buttons, and roadblocks that others have run up against. The library is a good starting point. The Internet and World Wide Web also provide access to a wealth of information. In addition, local universities and colleges have been implementing quality certificate programs for the past few years and now offer inexpensive evening and weekend courses as well as one- to two-day seminars taught by local consultants, registrars, and quality managers. These are all good places to start.

Join the American Society for Quality Control (ASQC). This organization promotes the theory and practice of quality control and has close ties with the American National Standards Institute (ANSI), the authority that controls the U.S. version of ISO 9000, known as the ANSI/ASQC Q9000. Membership opens up numerous channels that place you in the pipeline to receive current information about ISO 9000, from conferences and symposiums to opportunities to participate in local task groups and meetings. Jump in and attend those meetings. Network and talk to other companies who have undergone ISO 9000 registration. War stories abound.

Talk to your suppliers and subcontractors. Let them know that you will be implementing a Supplier Assessment Program as part of your ISO 9000 quality system and enlist their support in making your program work. Indeed, some of your suppliers may already be ISO 9000 certified and more than willing to assist your efforts and provide solid advice advantageous to both your quality systems.

If your organization has other divisions that have undergone ISO 9000 certification, invite them to one of your initial ISO steering committee meetings to discuss what they learned. If you're starting from scratch, invite representatives from other companies that have undergone ISO certification.

Learn as much as possible from others about what to expect. And while the specific contents of your quality manuals will be unique to your organization, you may be able to simplify the task somewhat by mimicking the style from an existing manual. Indeed, numerous software packages abound that provide templates to expedite the process. Discussing ISO requirements with others also helps you learn some of the pitfalls to avoid.

Ironically, learning from others also exposes the most frustrating aspect about ISO 9000 documentation: no two companies can or will approach it in exactly the same way. What often happens is that divisions within the same company will create different though equally viable solutions. For example, Varian's Oncology Systems unit, which is ISO certified in both the United Kingdom and the United States, has separate sets of manuals for each site. In turn, these manuals are completely different from those of Varian's Sample Preparation Product (VSPP) division where the emphasis is on assembly rather than manufacture. However, all of their manuals have a common format and closely aligned quality policy that derives from the corporate headquarters' emphasis on quality programs. Hewlett-Packard decentralized their ISO 9000 implementations in the same way, allowing each site to implement the standard independent of the others. In essence, while any organization can leverage the experience of different business units or operations, each must still must devise an individual approach.

A key area of leveraging the experience of others, and one which many organizations have used to expedite their ISO 9000 implementation, is the employment of ISO consultants and writers whose knowledge of and expertise interpreting the standard can smooth many of the issues you'll face during implementation. Consultants are often hired to provide guidance and to help kick start the ISO 9000 program. Another reason organizations use consultants is to bring in an outside perspective, often specifically seeking someone with specialized knowledge of their industry as well as ISO 9000 requirements. In addition, consultants can be very useful for reviewing the organization's procedures, providing an impartial critique (particularly helpful prior to an assessment), and offering an outside perspective. More details about the pros and cons of hiring a consultant are covered in Chap. 4.

Getting Top Management Commitment

All the advice in the world, either from this book or others, won't be of much help implementing the ISO 9000 standard without the commitment of upper management. A commitment that requires dollars, time, and resources as well as vocal, visual support. Many organizations have been unable to motivate staff successfully to put time and energy into documenting procedures and drafting a quality manual because there was not a corresponding demonstration of commitment at the executive level.

Ideally, this commitment should be in actions, not just words. It requires more than verbal assurances to motivate a company to the levels of quality most executives expect as a result of implementing ISO 9000. In fact, there is considerable consternation in the quality marketplace today that consultants have misled executive management regarding the productivity gains and cost-cutting opportunities to be obtained from implementing a quality program. It is not, however, the quality consultants who have erred. Rather it is management's unrealistic expectation that quality is a salve for a soft market or that employees will willingly implement a quality program which they think may endanger their job or increase their workload. ISO 9000 is not the holy grail of process improvement. Its implementation does not guarantee corporate nirvana.

What it does provide is an initial, necessary step on the path to total quality management that focuses on leveraging the internal customers' capabilities (i.e., the company's employees) and the external customer's expectations (i.e., the end users). There are companies who have made terrific gains in market share and profitability even in a soft market by addressing the true intent and spirit of a quality program; there are examples in this book. But such a move is bold and courageous, and cannot be launched by paying lip service to a quality program.

Lip service is better, however, than a blind eye. One of the keys, therefore, to getting the ISO 9000 program off the ground and to a good start is to gain upper-management commitment to provide the dollars to hire the necessary expertise as well as cover internal costs such as overtime, external training, and documentation requirements; to deliver the time required to review the upper-level quality policy manual and the performance of the quality system during various management quality review meetings; and to come through with the resources necessary to fulfill the program.

Perhaps it's a good idea to remember that executive management also needs occasional reenergizing in the form of success stories in order to remain enthused about projects such as the quality program. Don't keep the good news from the quality trenches a secret. Give them something to brag about at the next shareholder meeting.

3

Documenting Your Processes

The most critical point in pursuing ISO 9000 quality system registration is the documentation that supports compliance to the standard. Without documentation there is no objective proof that processes are consistently performed and appropriately approved. As a result, documents and records are the number one area to get scrutinized by external auditors. It is also the number one area where nonconformances occur (Fig. 3.1).

The process of documenting your processes therefore requires significant attention to ensure not only initial compliance to the standard, but the future ease involved with updating that documentation. Effectively defining the structure and content of the quality system documentation is critical.

There are numerous third-party commercial boilerplate and template software packages that hype the perfect solution to this documentation conundrum. There are numerous consultants and documentation organizations that claim to be able to document your processes in three months or less. Be wary of these excessive claims. Although certain portions of the documentation can be knockoffs of what everyone else is using, the reason your company has market share is because your internal processes are unique to your product, market, and company culture. There is no way any single company's processes can be shoehorned into a generic manual. Nor should you want to do so. Indeed, particularly in markets that operate on narrow profit margins, many companies consider their work processes to be their competitive edge. Such boilerplates or templates then are a starting point, no more. As with all starting points, you can use available tools to launch the process, but you will still have to go through the process in its entirety.

ISO 9001 Elements

Element		Percentage
Document Control	4.5	
Design Control	4.4	
Purchasing	4.6	
Inspection and Test	4.10	
Quality System	4.2	
Process Control	4.9	
Test Equipment	4.11	
Contract Review	4.3	
Corrective Action	4.14	
Management	4.1	
Quality Records	4.16	
Handling/Storage	4.15	
Internal Quality Audit	4.17	
Training	4.18	
Inspection/Test Status	4.12	
Product Control	4.13	
Traceability	4.8	
Servicing	4.19	
Statistics	4.20	
Purchased Product	4.7	

0%   5%   10%   15%   20%

Percentage

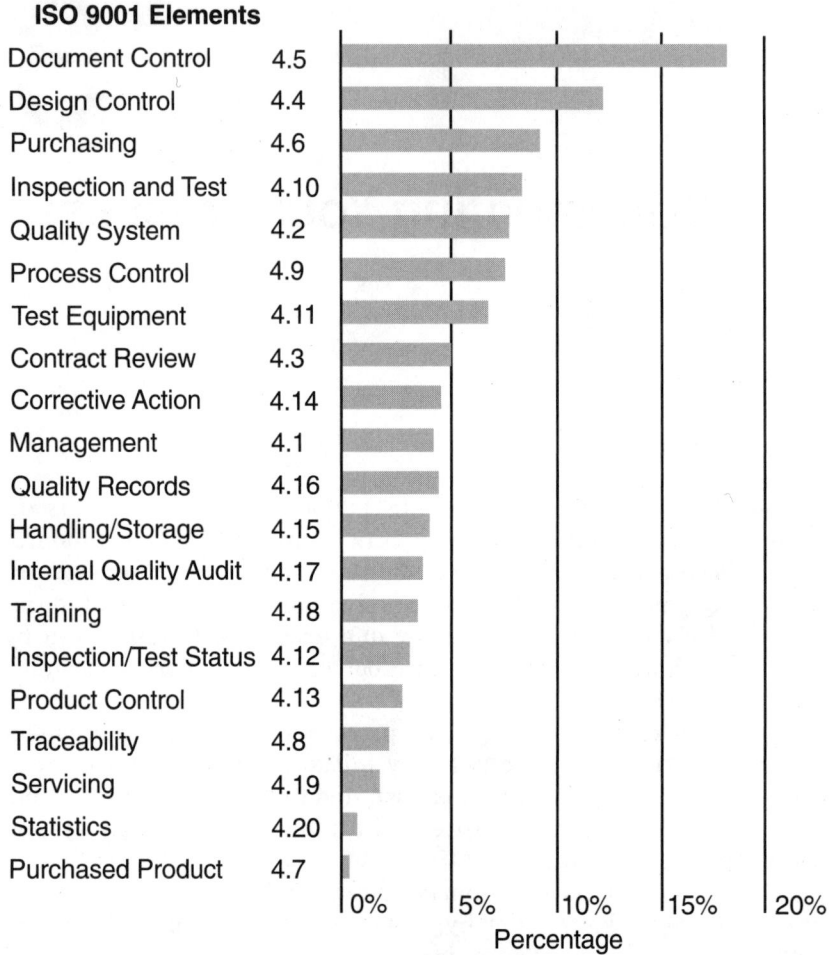

Figure 3.1 The percentage of system deficiencies registered against ISO 9001, based on 1040 nonconformances measured by Lloyd's Register of Quality Audits (LRQA).

The Bad News

If you run across claims that an organization can help you document and implement ISO 9000 in three months, you are being misled on four points.

First, no two companies perform the same process exactly the same way. The closest match to your quality manuals would probably be those of your competitors. The chances are slim to nonexistent that you will be able to access your competitor's ISO 9000 manuals to expedite your own documentation process. Even if you could, it might not have much impact on how you write your own documents. If you question the documentation house that is pitching a boilerplate manual, they will agree that additional processes must be written for specific cases, but they certainly won't have a written process specific to the step-by-step procedures that you will need to define. What you're really being told is that they have the boilerplate for a quality policy manual and can quickly guide you on how to write ISO 9000-specific processes that you probably don't currently perform, such as certain corrective and preventive action activities. What you're not being told is that because production processes are very specific, the boilerplate or template provided is simply a basic outline of key points. This basic outline would then need to be filled in with the appropriate details. Thus, there is no such thing as a generic manual that can be handed to a third-party auditor and enable your company to be certified to ISO 9000 post haste. The old adage, "If it looks too good to be true, it isn't," still holds true.

Second, while it appears to be relatively simple to write down what you do and follow that process, no one ever does. ISO 9000 documentation is dynamic. The moment a process gets committed to paper, a better way to perform the process is found and the process gets revised. Or, management takes one look and asks why the person or department is performing the process in such a roundabout way, and requests a modification. The modification must then become part of the normal operating procedures—something that doesn't happen overnight. Or, a bottleneck becomes painfully obvious and the process gets streamlined. Again, modifications require a learning curve. Or, one process is modified which then affects a second process and that second process must now be modified to ensure that both processes dovetail correctly and are not in conflict. There is a domino affect here. Hence, documenting your quality processes is just not very straightforward.

Third, outsiders cannot do all the work. You can't hire a corps of professionals to ram the ISO implementation project through. Those outsiders need to meet with many people in your company, taking up their time to ensure that the process is documented correctly. Time and again, your internal people forget or overlook steps—particularly steps that only happen occasionally—which should be documented. Many times these missing steps simply aren't apparent in one or two or more meetings. Expect several revisions for even the shortest procedures. And, there will be numerous internal discussions (arguments?) on what should get documented and why. Finally, everyone ignores the advice of outsiders if and when it suits them. There is no way to eliminate internal staff from involvement in the process nor would it be wise to do so because it is your staff, and not the outside professionals, who will be audited to the procedures. If your staff has not been intimately involved in the creation of those procedures, it will be extremely difficult to pass an external audit.

Fourth, even if by some miracle you managed to get all the documentation written in three months, no auditor in the world will certify you to ISO 9000. Why? Because ISO 9000 must be implemented. All those processes must be actively performed and the resulting documentation—the checklists, the sign-offs, the completed worklists, etc.—must be in existence. Certain criteria can be grandfathered in, but at some point you must start a paper trail that confirms adherence to the ISO 9000 processes, and the auditor will expect to see a reasonable amount of proof—generally at least three months worth of records. So a minimum of six months is required, right? Wrong. In those six months, you must also perform internal audits. It's almost a given that your internal auditors will discover many nonconformances which, as a conscientious company, you will want to fix before you call in an external auditor.

This is why most companies spend eighteen months or more implementing ISO 9000 to the level necessary to pass the initial third-party audit. These organizations are not dragging their feet. Such an effort requires department meetings, interdepartment meetings, upper management meetings, quality team meetings, internal quality audits, revisions and rewrites, training, and more. What the company markets cannot come to a standstill while ISO 9000 gets implemented. In fact, the big joke in the late 1980s was that many companies who achieved the Malcolm Baldrige National Quality Award subsequently went out of business because they spent too much time on their quality system and not enough on the business of being in business.

The key is to take the time to do it right. Give the labor-intensive documentation process the same care you give to the processes you perform to produce your organization's product or service. ISO 9000 will redefine the underlying structure of how your company operates. It will drive numerous significant decisions regarding how the business is run, as well as how the product is developed and produced. In addition, the trends toward increasing government regulation in all aspects of business will require a robust, thoughtfully developed, comprehensive quality system. With so much at stake, a snap decision just won't do.

The Good News

Now that you have the bad news about the amount of time and effort required to develop appropriate documentation, the good news is that it is a relatively straightforward process despite its time-consuming nature. Many companies have used their ISO 9000 documentation to ferret out process bottlenecks, streamline tasks, and significantly improve productivity. What makes documenting your processes difficult is not so much the performance of the task, but the long-range nature of the work. As with any project, break it down into components, strive not to place too great a burden on any particular department or person (with the exception of the ISO Project Leader and his or her assistants), and keep plugging away until it gets done.

The Four Levels of Documentation

There are typically four levels of documentation developed to address ISO 9000 quality system requirements (Fig. 3.2). The top level is the quality policy manual, which contains policies that address each of the requirements. The second level consists of the quality procedures, which contain procedures that define activities and responsibilities. The third level consists of the quality work instructions, which are the step-by-step activities. The fourth and final level is composed of miscellaneous support documents such as specifications, standard operating procedures (SOPs), appendices, and log books.

Many companies choose to position the second and third levels as implementing procedures and operating procedures, but what gets covered in the two levels is basically the same no matter what they

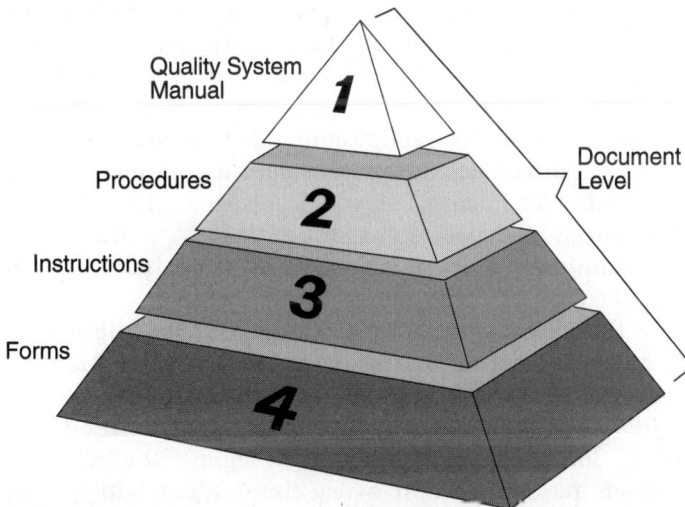

Quality System Manual **1**

Procedures **2**

Document Level

Instructions **3**

Forms **4**

Figure 3.2 The documentation pyramid sets off the various segments of the quality manual documentation, from the highest policy level to the lowest records level.

are called. In the second level you define the procedure, and in the third level you spell out the steps.

There are three ways to approach the documentation process: implementation by clause, implementation by department, or implementation by process. Each approach has its merits; all are acceptable.

One of the most popular ways to document a company's quality processes is based upon the clauses or elements of the standard. In this approach, the documentation typically uses a similar numbering sequence as that used by the ISO 9000 standard, grouping the procedures and work instructions according to the various elements in the standard.

Many companies choose to organize their procedures around functional areas such as order processing, shipping, receiving, and manufacturing. Implementation by department or function makes it easy to locate procedures quickly, but this runs the risks of procedure duplication and inconsistent handling of procedures that are companywide in scope, such as corrective and preventive action.

Implementation by process is less common and each procedure or work instruction follows a single process from start to finish. In this case, criteria such as corrective and preventive action are not separated out. Rather, the performance of those activities is embedded within the particular process.

Quality policy manual

The quality policy manual is the one section of the ISO 9000 documentation that can effectively use a boilerplate manual, and in many ways it is wise to do so. First, most such boilerplates mirror the language in the ISO 9000 standard. Second, most third-party auditors expect to see the elements of the standard covered in the quality policy manual. A boilerplate quality policy manual, however, certainly doesn't require a high-priced consultant to construct. Many software templates are available for about $100 which can focus the organization on what's required to expedite the initial draft. A written template for such a sample quality policy manual is included in the appendices to this book.

When starting the ISO documentation process, many companies find they don't have an existing quality policy, much less a manual, having previously relied upon a general requirement to "do the job right." ISO 9000 requires that such statements get defined and fleshed out. What does it mean to do the job right? How will you ensure that the job gets done right every time? What will you do when the job doesn't get done right? Your quality policy manual sets up the initial response by providing a commitment by the organization to address these questions with specific actions. This is one of the first documents to be developed, and the first to get delivered into the hands of the third-party auditor for inspection.

Most organizations base their quality policy manual on the 20 elements of the standard as described previously. This ensures that all elements of the standard are covered, that the organization has committed to addressing the required criteria, and that the information is presented in a format which is obvious to both the organization and the auditor. It is simple and quick to modify a generic manual to your company's needs.

As can be seen by the template provided in Appendix B, the scope of each element must be defined and the policy concerning each element must be laid out, as well as a basic list of the parties responsible for ensuring that the particular policy is enforced. The lower-level documentation in the quality procedures and quality work instructions defines how the quality policy is put into action.

Generally, a quality policy manual is about 25 to 35 pages in length. The main point to keep in mind is that it needs to demonstrate commitment to the elements in the ISO 9000 quality standard.

Quality procedures

The quality procedures are the second level in the quality documentation and are a means of controlling, identifying, and specifying the activities that produce a product or service. In most cases, interdepartmental procedures that address consistency and quality issues—such as document control, calibration, and internal quality audits—don't exist prior to implementing ISO 9000.

Quality procedures typically outline department and individual responsibilities, and point to the specific quality work instructions that each of those departments and individuals perform. If this level of documentation is known as implementing procedures, the contents are roughly the same and also address who is responsible for implementing the procedure whether it is a department or a person.

Interdepartment responsibilities are often defined in the quality procedures. This enables both the organization and the third-party auditor to determine where one department's responsibilities end and another department's responsibilities begin. This ensures that the handoff of paperwork and products between one department and the next gets handled consistently.

When individual responsibilities are being highlighted, the position title—not a specific person's name—is used. Some companies, particularly European organizations, include job descriptions in this area. However, this is not always a satisfactory solution. A job description usually contains many duties not relevant to ISO 9000 or a quality system. Including a job description would then include additional text that the organization could be audited against (anything in the quality manuals is subject to audit), but which are not covered elsewhere. If the person is performing tasks not defined in the job description (as is usually the case in the United States), then the person is not in conformance to the standard, and the auditor could write a nonconformance to that effect.

It is much better to define only the responsibilities and approval authority specific to the function being defined and the requirements of the ISO 9000 standard.

The quality procedures should also provide a list of references and related sections, providing a cross-reference to related procedures and work instructions. For instance, a quality procedure on Document and Data Control may refer to quality work instructions for Procedure Control, Document Maintenance, and Record Retention. Each of those criteria is usually covered separately. Procedure Control, for example, specifically applies to how the ISO documentation is controlled. Document Maintenance, on the other hand, applies to how working documents such as orders are to be maintained. Record Retention typically defines the length of time different documents and records are to be stored and is often provided in a matrix format.

Quality procedures also often contain a short section that defines the responsibilities pertinent to the records appropriate to the procedure.

An outline for such quality procedures may contain the following sections:

1. Purpose

2. Scope

3. References/related sections

4. Interdepartment responsibilities

5. Responsibilities and approval authority

6. Records

Or, it could be organized in the following way:

1. Purpose/scope

2. References/applicable documents

3. Definitions

4. Responsibilities and approval authority

An Implementing Procedure might be organized along these criteria:

1. Purpose

2. Scope

3. Terms and definitions

4. Responsibilities and authority

5. Reference documents and standards

6. Records and forms

7. General procedure

As is apparent, certain elements are standard for such quality documents, while others can either be addressed in this level or in related appendices, which would then be listed in the related documents or references section.

Quality work instructions

The real detail and bulk associated with ISO 9000 documentation resides in the quality work instructions. Also known as operating procedures, the quality work instructions define the operational steps taken to perform specific procedures. There are a couple of ways to organize quality work instructions.

One of the most popular is to present the quality work instructions such that they roughly follow the process of what occurs when an order is received, how it is subsequently filled, and then finally shipped to the customer. Each key procedure is written separately. If you look at the ISO 9000 standard, the standard itself follows this process roughly. Purchasing is placed in the beginning and Training at the end, but Contract Review (which includes order processing) comes before Process Control (which defines manufacturing processes), which in turn is placed in front of the requirement for documenting Handling, Storage, Packaging, Preservation, and Delivery processes.

Generally, each quality work instruction has a short introduction that defines its scope, followed by the step-by-step activities. Some organizations choose to include a list of definitions for certain industry-specific or company-specific terms. These definitions could also be placed in the quality procedures or in a separate appendix to your quality manuals.

The quality work instructions should include all the peripheral activities that go along with performing the actual steps in the particular process. This includes

- Prioritizing or organizing materials or paperwork for the process
- Preprocess checks
- What to do if there aren't enough materials to fill the order
- What to do if the order is canceled while in process

- What to do if nonconforming product is discovered
- What to do when the order or batch is complete
- Any special processes required for certain customer or product orders

These requirements can be provided in a single work instruction or broken out and provided in separate work instructions. The choice you make depends upon the process being documented and how long the resulting work instruction would be. For the sake of document administration later on, it is easier to write many short work instructions than one long one. If only a short section of the work instruction is modified, you certainly don't want to reissue a lengthy document when you can reissue one that's short. The paper costs alone make this nonsensical. In addition, most people don't like (and don't have time) to wade through long documents. Your goal is to get them to read and understand the material. Make it easy for your staff to do so. Write short procedures and work instructions that are easy to comprehend, easy to use, and easy to update.

Specifications, SOPs, appendices, and log books

There are many supporting documents within any company. These can be the SOPs, lists of product specifications, a reference binder of all quality-related forms, or log books that support activities such as instrument calibration. These must all be kept up to date and accurate because they too support the activities surrounding the quality system. This documentation is also subject to review by the third-party auditor.

Interpreting and Documenting Processes

The volume of documentation that must be formalized can be daunting. If you have defined the documentation structure and assessed your existing quality system, then documenting your processes becomes a pragmatic, albeit time-consuming, project.

When reviewing existing documentation, you'll discover one of the following:

- The documented procedure exists and is correct.
- The documented procedure exists and is incorrect.
- The documented procedure exists but is not followed.
- The procedure is followed but not documented.

Most of the time, procedures just need to be written down and formalized. But certain activities and functions, such as approval matrices and a document control center, may need to be implemented based upon the ISO 9000 requirements.

Often not rushing into documenting your processes can increase the quality of the end result. In addition, allowing the various departments to implement the standard at their own speed develops ownership. Provide a deadline, of course, and monitor progress, but allow staff to set their own pace.

Let people come at it slowly. Give them the tools to perform the task, such as outlines, and let the various quality project teams document the parts of the process bit by bit. Meet weekly to discuss progress. Once a first draft exists, address the specifics that are required by ISO 9000.

To simplify the process as much as possible, many companies use flowcharts, both to define the process initially and as part of the final documented procedure or work instruction. For organizations that employ multilingual staff, particularly on the production floor, the visual nature of a flowchart makes it easier to remember and follow the procedure as defined. It also makes it easier for these staff to answer an auditor's questions. Some companies even use photographs to document work instructions.

"During the course of implementation, we put together several new procedures, including management review and internal audits, as well as formalized our customer complaint procedure," remarks Rob Ireland, Quality and Regulatory Manager, Beckman Instruments Ltd in the UK. "In service, there had been procedures in place for installation, training, and servicing, but the warehouse had nothing down in writing per se. Our training program was well established and became much stronger with formalized on-the-job training strategies. For instance, we found one of the dangers with on-the-job training is that it repeats bad habits, so we keep a close eye on this aspect of the program."

Beckman's three levels of documents have "extensive cross-referencing and sign-offs by affected departments at the end of each procedure," Ireland adds. "Each department has a complete set of manuals designed for accessibility between departments so they can refer to their own and others' procedures. But certain information is password-controlled on computer. We use a combination of computer and hard copy to ensure marketing and proprietary information remain confidential."

In order to keep updates from becoming an administrative nightmare, it is wise to keep procedures and work instructions as short as possible—no more than a few pages. Indeed, some organizations only use flowcharts, leaving the details to be covered by training. This way, the burden of proof rests with the training program, thereby eliminating excessive detail from the written procedures.

This brings up a key point—don't overdocument. Keep your documentation simple. Do this by keeping an eye on the volume and level of detail. Often, when the people who perform the procedures are asked to write the procedures, they write down every tiny detail. Instead of just writing "soldering," they write the steps for how to perform soldering, such as "turn pin over joint." Unfortunately, this makes the quality manuals too detailed and includes things that are peripheral to the actual process being defined. After all, is the purpose of the work instruction to tell the person how to perform soldering or how to perform product assembly? The main thing is to exclude nonessential issues, such as how to fill out forms or how to solder, from your procedures and work instructions.

Writing concise procedures and work instructions requires a reciprocal beefing up of training. The more training a person receives, the less he or she needs to rely on written instructions to ensure that a process is performed consistently. Your organization will need to define the balance between training and work instructions, because how much or how little training you provide your staff will dictate how much or how little documentation you need to develop.

Remember that your company will be audited to the documentation, and excessive documentation makes it difficult to pass an audit. If one person performs even one step differently from the next, a nonconformance is inevitable even if the difference in how the procedure is performed has no impact on the end product. ISO 9000 is not concerned with whether your organization turns out a quality product; it zeros in on whether your organization consistently follows the quality system documentation. So, don't allow your company to be trapped into excess documentation.

Here are a few more reasons to keep your documentation short and to the point:

- Simple documents enable new employees to learn quickly.

- The documentation will be current longer since fewer changes and updates will be required.

- The documentation will not be open to interpretation by the auditors.

In summary, you'll need to develop a standard format; draft the top-level quality manual; revise existing procedures and work instructions; and create new procedures and work instructions. Once you've determined what processes you need to document, you'll need to determine any additional steps required to address ISO 9000. Appendix A provides an element-by-element guide to interpreting the ISO 9000 requirements.

Remember that because ISO 9000 is a voluntary standard, there is no set format nor does every company create the four levels highlighted above. Sometimes procedures and instructions are a single document. The key is to develop a documented quality system that works for you.

Implementing Your Processes

Surprisingly, one of the biggest stumbling blocks to achieving ISO 9000 certification is one that seems so obvious that no one could overlook it. But many organizations have. This stumbling block is implementing the processes that have been documented. If you write down that everyone fills out a form in a certain way, but you haven't trained people to use the form, and the form is not consistently filed for easy reference, then you have not implemented the process.

It's amazing how many companies state that they have a simple operation and they intend to apply for certification within three to six months of when they first start documenting their procedures. It's a nice idea, but virtually impossible. The organization has no track record for the documented process. Without a track record, there is no history of how well the process is performed, and thus no history that the process has been performed according to the documentation. Therefore, a certain amount of time must pass before the organization is ready for a third-party audit. Before that time several things must happen.

Closing the Loops

First, you must review each process as it is performed against the documentation to ensure that all handoffs between departments are completed appropriately and that all loops are closed. No document or process can be allowed to be a one-way street with no closure.

To ensure that closure happens, someone must review all the documentation and cross-check the handoffs between departments. Often this person is the ISO Project Leader or the Quality Assurance Manager. Sometimes the task is assigned to someone in document control. When reviewing the documents, look for statements that say "the form is forwarded to Inventory." Then look at the Inventory procedures or work instructions to find out what they do when they receive the form. If the form is not mentioned, add the appropriate steps to state who receives the form and what that person does with it.

Sometimes forms are not involved. Sometimes a department is notified verbally for assistance—particularly when a nonconformance is discovered that needs an immediate fix. For instance, if a nonconformance is discovered on the assembly line, and the line is stopped, and someone from another department is notified to check the nonconformance, then that other department must also mention what happens when they are notified of such incidents.

Closing the loop, therefore, means scrutinizing your quality manuals for actions and paperwork that cross between departments or functions, and then checking the referenced department or function to ensure that the action or paperwork is also covered in their procedures or work instructions.

On the other hand, if you decide to keep all actions within the same procedure or work instruction, then you must ensure that all steps taken by the various responsible persons are described. As mentioned in the previous chapter, there is a danger to this approach. The Inventory Clerk may say that because it is not an Inventory procedure, he or she did not know about it. This is not an acceptable answer, of course, but it points out one of the difficulties of documenting a large number of processes—that departments tend to look at their own department procedures and no one else's. After all, most ISO quality manuals are large enough to discourage anyone from reading more than is necessary, and training someone to know where their processes are throughout the manuals is difficult enough. Because ISO documentation is dynamic and most organizations are in constant flux, what Inventory did one quarter may be the responsibility of the Warehouse the next. Even so, it makes sense to organize documents by function but reference other departments. This actually simplifies closing the loop between procedures since you can quickly reference the department mentioned in the handoff.

Training

Once the documentation is complete and all the loops and handoffs are closed and resolved, then staff will need to be trained—or retrained as the case may be—to the new quality manual criteria. You may think that everyone is performing a process the same way, but they seldom are—there is always some variation. Variety may be the spice of life, but it won't do when you are trying to measure processes using statistical techniques, nor will an auditor believe that your processes are in control if everyone performs the same task a little differently. Training, therefore, is meant to ensure that everyone knows what they are expected to do according to the quality manual, that they know where to locate their procedures within that manual, and that they understand any changes to their procedures which have been implemented as a result of ISO 9000.

This training must, of course, be documented and filed in the appropriate training records. As time goes on after your company receives ISO 9000 certification and is happily promoting itself as an ISO 9000–registered company, you will find that you still have nonconformances. This is because the nature and type of nonconformances will continually change. What will happen over time is that those nonconformances will not necessarily be due to poor-quality raw materials or to out-of-calibration equipment. The nonconformances will be caused by the people checking those raw materials, using and monitoring the equipment, and filling out the forms required by the procedure. If the raw materials are faulty, someone let them in the door without correctly following the incoming inspection procedure. If the equipment is out of calibration, someone did not calibrate it according to the calibration schedule. If the forms and checklists are not filled out, the person filling out the form is at fault, not the form. Over time, people most often become one of the biggest causes of nonconformances, not product problems. Hence you'll find that a significant portion of your response to auditor queries for the cause of nonconformance will be operator error, and that your fix will be retraining.

To err is human, of course, but this doesn't make the problem of resolving nonconformances any easier. Perhaps more than anything else, this issue of constant training, that is, constant reminders that ISO is to be in the top of employees' minds whenever they perform any task, underscores the importance of actively involving staff with the ISO 9000 process. Many companies do this by requiring the people who perform the process to write the process. This is fine if the person has a good grasp of English, but many multicultural workplaces employ perfectly competent workers whose writing and reading skills are elementary at best. In addition, many times when the person who performs the task is given the task of writing down the process, he or she includes every possible step in the process down to irrelevant details—details that do not need to be spelled out and which will only hamper an audit. These two solutions do not work for every company. What does work is training.

Typically, training is performed classroom style, with one person talking the group through the key points of the procedures and discussing what has changed and why. The "why" is very important here, particularly if management has decided that an old step is no longer necessary or that a new step is now required. The better people understand why they should perform a step a certain way, the more likely they are to do it the way you want, that is, the way it is written.

In the case of retraining someone who has been performing a procedure incorrectly, and one or more nonconformances exist to prove it, one-on-one training is usually the better approach.

When training doesn't work, and the same person continually repeats the same nonconformance, you need to do two things. First, examine the process. Is the process written so that performing it correctly is extremely difficult if not impossible? Training will not be able to solve a faulty process. Such a review of the process itself should also be part of your Corrective/Preventive Action procedure, which seeks to determine the root cause of nonconformances. Second, if extraneous factors are not the problem, but the person performing the procedure simply can't perform it correctly, then you have to make the difficult decision either to reassign that person elsewhere, or place the person on probation. If you run into such a problem, remember that in W. Edward Deming's 14 points, he stresses driving out fear so that everyone may work effectively for the company. Perhaps the problem goes deeper than the process or the person's abilities, and lies a misconception that ISO 9000 requires unnecessary work or steps. If this is the case, no matter how much you exhort the employee, that person never sees the value of supporting the quality management system. Conforming to ISO 9000 requires extra work that few people are delighted to perform. Performing those tasks will create a cultural change. Change is never easy and usually resisted by some. Allowing those who resist the change to continually undermine the system will ensure that the system you are attempting to put in place won't work right.

Objective Evidence

After you've documented a process, cross-checked the documentation, and trained everyone in the new quality procedure requirements, you need to ensure there is objective evidence that people are doing what the documents say they are doing.

Objective evidence is made up of many elements. In the case of training records, what if the job requires specific external training? For instance, warehouse workers may need to be licensed to drive forklifts (you would be amazed at the number of forklift accidents that routinely happen in the average warehouse—just go look at the doorways). To protect product and prevent accidents, many companies require specific driver training before letting warehouse workers drive forklifts. When this is the case, there must be some kind of documented evidence, such as a copy of the certificate of completion, which shows the person has been qualified to drive the forklifts.

Objective evidence can also be as simple as a signature and date on a release form which proves that a product was approved for shipment. That form then needs to be filed in an appropriate place for a specified amount of time.

Objective evidence can also be the calibration log books or tags associated with a particular piece of equipment. If that log book or tag is initialed and dated when the equipment was last calibrated, then you have objective evidence that you are following the written procedure.

In all three cases, objective evidence consists of some form of documentation which proves a task was performed as and when required. It is this objective evidence that the third-party auditor will ask to see when auditing your processes. The documentation describes how the task should be performed, and the objective evidence proves it was done correctly.

Many companies satisfy the requirement for objective evidence by creating checklists that the person ticks off and signs at the end of a process. This is a great solution to minimizing the amount of documentation that must be used to satisfy the ISO 9000 appetite for objective evidence. Keep it simple, right? But, if you don't have checklists and decide to create them, don't forget that you are changing the process and that you will need to ensure training is provided to all people expected to use those checklists.

If it begins to seem as if you can't make one move without making another, you're right. ISO 9000 lets you change your processes any time you want, any way you want. But you must ensure that the change is followed through to the extent that everyone involved knows about the change, can perform the changes, and is trained appropriately. Initially, these requirements may seem like excessive paperwork and effort on your part. This is not the case. Rather, by ensuring that these steps occur, you will have greater chance of ensuring that the change is commonly understood and followed throughout the organization. And, you will know that the change is being performed correctly because there is objective evidence to prove it.

Document Control Issues

The collection of objective evidence in turn requires control of the objective evidence. Most organizations handle this by creating a Records Retention procedure and matrix that details how long records, forms, procedures, and other quality-related documents are to be kept and where. Typically the department that initiated the form is responsible for maintaining it. When specifying how long a record is kept, many companies take into consideration legal requirements or statutes of limitations, and incorporate those requirements into the document or record retention period.

Whenever a document is a controlled quality document, not only must that document be stored appropriately, but the length of time it is to be kept must be specified. For instance, you may want to keep records of your customer orders and contracts for four years after shipment before disposing of the document. Or, you may want to keep all training records for three years after employment is terminated. If the record is product-related, such as the Approved Vendor List, you may want to keep copies of that document for a certain period of time after shipment of product or receipt of the raw materials that went into the product.

When writing your document control procedures, there are three key points to keep in mind. First, you need to specify the filing system. In other words, is it the responsibility of the department that initiated a form to maintain it? If so, state that clearly. Second, how will your organization withdraw records and documents from the files or archives? When someone wants to look at an archived document, you need to ensure that the document gets put back into the archive when they're done with it. Third, what will you do with out-of-date documents? To do this, you need to specify how documents will be reviewed for disposal, who will perform the review, and how often that review will occur. If out-of-date records and documents are to be destroyed, you may want to have a second level of approval before this can happen. After all, you can keep a record longer than the specified retention time (if you put this clause in your documented procedure), but you can't retrieve a document that no longer exists.

Don't forget that if you archive documents outside the department, such as in a warehouse (and you will eventually need to do so), you will need some kind of Quality Records Archive Location List which states where those documents are stored.

Conducting Internal Audits

Almost without exception companies need to implement an internal audit procedure. Very few organizations have one in place. The procedure for internal audits should include assigning internal auditors to scrutinize processes, checking that documents are filled out and maintained appropriately, and ensuring that the person performing a task has been trained to do so. Selecting internal auditors and determining how many you should have can be a creative and innovative process. Many companies, for instance, use a consultant to help perform the initial internal audits.

The number of auditors you have will depend not only on the size of your organization, and on the number of departments involved, but on how much time those auditors can direct to their ISO responsibilities. One solution is to create a core team with additional backup auditors to take the burden off the core team. Using a large number of auditors gives more people a better understanding of the business and improves communications. In addition, drawing from a large pool of auditors keeps the auditors from appearing to the rest of the company as "ISO Police." If many different people audit a department or function, then it is hard to target the auditor as someone looking to find something wrong. Alternatively, some companies use their partnerships with other companies to trade internal auditors in order to get a fresh pair of eyes at no cost.

One thing that is important about internal auditors is that they must be objective and shouldn't audit the area in which they normally function. It's important to select and assign internal auditors so that they audit an area different from where they work. The standard requires this, but it goes without saying that an organization shouldn't have engineers audit engineering. Also, if you use a variety of auditors, it's best to set up a rotating schedule so that staff aren't overburdened.

Who do you choose? Select people who are articulate, amenable, have a good understanding of the business, and who are well-respected by their peers. Internal auditing is a great cross-training opportunity, and one that often promotes cross-pollination of good ideas and procedures between departments. People start to recognize the entire process and help each other. It gets people more actively involved in the overall goals of the process.

Determining the number of internal audits to perform depends on the intensity level you choose to assign to the auditing process. The fewer the audits, the more extensive each must be. Most organizations opt for quarterly audits of all departments. Even with this level of frequency, you may find that significant changes take place even in just three months. If the company is particularly dynamic, the organization chart will change slightly all the time. And if you include the organization chart in the quality policy manual, then you will know of at least one page that always needs updating and auditing. One solution is to devise a matrix showing when an area is to be audited and who will perform the audit. In fact, the external auditor will expect to see a document like this. This Internal Audit Schedule should show the planned internal audits that will be performed on an annual basis, as well as indicate whether the audit took place as planned.

To document the results of an internal audit, many companies use a form very similar to the form used by the external third-party auditor.

This makes perfect sense. After all, if you use a similar form, then you will be putting yourself in the same mind-set as the external auditor and thus be better able to find the same types of errors that the external auditor would find. When this happens, you ensure the likelihood of fewer nonconformances because your internal audit team will have already unearthed the majority of problems. Once uncovered, those problems can be addressed through your Corrective Action procedure, significantly reducing the possibility that the same issue will be the cause for a nonconformance recorded by the external auditor.

To perform an internal audit, the internal auditors will need to work from a checklist of what to look for and write up any nonconformances that are discovered on a Nonconformance Report. The checklist should be based upon the ISO 9000 standard and should be used as a prompt for the auditors to ask the appropriate questions regarding the function or department they are auditing. For instance, if the internal auditor is assessing the Purchasing function, requesting to see the Approved Vendor List would be expected. The next question might be to see the purchase order for the last shipment from a specific vendor or subcontractor. Then the Incoming Inspection Checklist for that shipment may be requested. With each document, the internal auditor is checking to see that the form or document is filled out correctly and signed and dated by the appropriate person. If any inconsistencies are discovered, a nonconformance will need to be reported. For a look at a sample Nonconformance Report, refer to Appendix C.

Nonconformances are reported on the Nonconformance Report form, which should include the following components:

- The Nonconformance Report number (assigned each time one is written)
- The name of the company being audited
- The name of the department or area being audited
- The date of the audit
- The name(s) of the person(s) who was audited and his or her title(s)
- The name of the auditor
- The nonconformance observed, whether it is major or minor, and what clause of the standard applies
- The cause identification and proposed corrective action
- The agreed time period for implementation of the corrective action
- The name of the person responsible for the corrective action
- The department supervisor's signature agreeing to the nonconformance and time to fix

- The auditor's signature
- Finally, a section to be completed later that addresses whether and when the corrective action was successfully completed within the specified time frame, including a signature and date by the appropriate authority.

In addition to the Nonconformance Report, the auditor may also use a Quality Assessment Report to highlight any observances. Observances are generally made regarding processes that are not being performed as well as they could and which the auditor will review in closer detail during the next audit. Observations should be treated as warnings. For instance, a more formalized approach for incoming inspections may be suggested to better track vendor product quality and performance. It is wise to take the time to review your processes and modify the procedure accordingly before the next audit. A sample Quality Assessment Report is included in Appendix D.

It's not unusual to be three-quarters of the way through the ISO 9000 implementation process only to discover there are staff who are performing a procedure incorrectly because they still don't understand what ISO is about and are not using the system. Internal audits can develop ownership of the processes that the person performs. It's interesting that internal audits can also create peer pressure and competition. No one wants to belong to the department that doesn't pass an audit.

The Pros and Cons of Using Consultants

Many companies use the services of a consultant to help with various aspects of implementing ISO 9000. This could include hiring a consultant to help direct and organize management during the initial stages of setting up the ISO 9000 quality management system. It could include using a consultant as a member of the ISO 9000 steering committee to identify procedures below the quality manual. It could include bringing in a consultant to perform training. It could include using a company that specializes in ISO documentation to write your quality manuals. Or, it could include using a consultant to help perform internal audits. Generally, organizations use consultants to help clarify and expedite the ISO 9000 implementation process because the consultant is already well-versed in understanding and applying the standard's requirements.

There's little doubt that a good consultant can ease and expedite implementing ISO 9000. You can bring a consultant in at any time during the implementation process, depending on what type of consultant you wish to use and what you need the consultant to do.

However, finding a consultant you can count on is easier said than done. There are horror stories being told by companies who have spent thousands of dollars on consultants and external documentation houses that have resulted in incorrect information and inadequate manuals. Worse, sometimes those companies are no further along in the ISO implementation process than they were when they first hired the outside assistance. But if the expertise is not in-house and your organization doesn't have the time or personnel to spare, hiring a consultant is often the only recourse. So how should you go about finding a consultant who will meet your needs?

You will need to do a bit of research to discover and locate a good consultant. One tactic is to attend local ASQC (American Society for Quality Control) meetings. Many consultants go to those meetings; indeed, your local chapter may even have a list of consultants for you to call. Attend some of the quality congresses and trade shows. The ASQC hosts numerous events around the United States every year; there are just as many being held in Europe and elsewhere in the world. Consultants are called upon often to provide presentations at these shows. Introduce yourself to the ones who impress you. Finally, call upon your own network of peers. Ask for referrals from other companies who have implemented ISO 9000 and used consultants.

Once you have a list or found a consultant you like, the key point to remember is that expertise, not education, is the most important aspect. That's because there are no ISO degrees. There are no minimum education requirements. Anyone who wants to can say he or she is an ISO consultant. Many people who have been involved in quality control for years and have earned a CQE (Certified Quality Engineer) designation from the ASQC are promoting themselves as ISO consultants. While those persons are intimately involved in quality issues, a CQE certificate does not mean the person thoroughly understands ISO 9000. Many outstanding ISO 9000 consultants do not have any specific quality certificates. Instead, a good ISO 9000 consultant's track record will speak for itself.

Hiring a consultant to kick start your ISO 9000 program should be approached in the same way as hiring a consultant for any other task. Obtain a detailed checklist of what the consultant proposes to do for your organization and what you'll expect to know or have at the end of the relationship. Get a price quote: keep in mind that cheapest is not necessarily the best option, nor are the most expensive always worth the money.

When interviewing a consultant, find out about the specific ISO 9000–related experience the consultant possesses. Ask how many companies the consultant has helped achieve registration to the standard. Find out if the consultant is familiar with your industry and understands the idiosyncrasies of your market's processes and requirements. Ask for references from other companies that the consultant has helped through the ISO 9000 process, and call those references.

When discussing the pros and cons of using an ISO 9000 consultant, the response from ISO-certified organizations is very mixed. Some organizations think that a consultant is a complete waste of time and money. Others were delighted with the assistance the consultant provided. The key seems to be in whether the consultant who was hired was right for the task. For this, you need to be clear as to exactly what kind of consultant you want. Do you want an outside authority to train upper management or train internal auditors? If you want to train internal auditors, it may be better to sign up the internal auditors for a local university course on the subject. Indeed, you may even want your ISO Project Leader to take Lead Assessor training.

"We interviewed five consultants," explains Thermo Separation's Quality Systems Manager Ron Haines. "We looked for good presentation skills and a strong quality knowledge, then cost and experience, particularly how much actual experience they had bringing someone through an audit."

Consultants are not necessary to a successful implementation; however, one that's a good fit with your company can be invaluable. The trade-off is in time versus money (isn't everything?). Certainly consultants can speed up the process, but their cost also makes the project more expensive.

The most successful employment of consultants appears to be in how they are utilized by the organization. "While we used consultants to orient and educate staff on what to expect," recalls David Lowe, Varian Analytical Instruments' U.K. Country Manager. "the first one we hired led us down the wrong path. The second one was much more effective— an ex-British Standards Institute person who lectures on quality issues at local colleges." Varian learned the hard way, although they were ultimately successful in their search for an effective consultant. "I highly recommend asking how many companies the consultant has helped through ISO registration before you hire anyone," Lowe adds.

Thermo Separations also used consultants for educational purposes. However, Haines cautions that "when you allow the consultant to do most of the work, then you're left in the dark because you really don't understand the standard. So we did the implementation plan ourselves and used the consultants for training."

Beckman Instruments' Quality & Regulatory Manager Rob Ireland states that "early on in the process, we talked to a lot of consultants and decided to use a consultant initially to kick things off, to help get documents going, and to ensure that we addressed issues in the standard. During the selection process, we asked who the consultant had helped to reach ISO registration, how many companies he had helped, and what registration body he used."

Another reason organizations use consultants is to bring in an outside perspective, often specifically seeking someone with specialized knowledge of their industry as well as ISO 9000 requirements. In addition, consultants can be very useful for reviewing the organization's procedures, providing an impartial evaluation (particularly helpful prior to an external assessment), and offering an outside perspective.

The same methodology for selecting a consultant can be used to select a freelance ISO 9000 documentation writer. One of the major obstacles to ISO 9000 implementation is the time required to document procedures and work instructions. Often these tasks are assigned to already overworked managers or supervisors. Consistency of style or terminology cannot be expected. Often these writing tasks are delegated to the staff who actually perform the function. Unfortunately, clear writing skills may not be their strong point and they may misinterpret the ISO objective by writing a procedure that their manager wants to see rather than one that reflects the actual process. The choice is whether to undergo the often painful documentation process in-house or use an external writer to interview staff, observe the actual process, critique any written materials, and assist the organization in developing a consistent set of documentation that interlocks across all processes. Again, experience is key.

Alternatively, you may want to consider using an ISO documentation house that specializes in writing ISO documentation. In this case, don't believe all the documentation promises you hear. There are ISO documentation companies that promise to write your manuals in a few weeks, making you ready for an audit in a short period of time. If this sounds too good to be true, it often is. There are instances where such an organization was hired to write ISO documentation only to present inadequate results and be replaced by a single freelance ISO documentation specialist. The ISO documentation specialist then worked closely with the various departments to customize the manuals to that company's unique set of processes. Whether using a single dedicated writer or a team of writers, it is important for the writer to meet with the people performing the tasks and to write the procedures accordingly. Despite these caveats, there are reputable ISO documentation houses that perform an admirable job—some of them even assemble a cadre of freelance specialists with expertise in your industry to perform the task. The important issue is to do your homework when selecting documentation specialists just as you would with any consultant or outside contractor.

A word of caution about hiring outside ISO writing assistance: you get what you pay for. Many ISO documentation companies use a generic template that gets modified for your organization's processes. This is fine in principle, but in reality it often means that they are trying to shoehorn your company's processes into an ill-fitting format. No one company performs their processes exactly the same as any other company—even direct competitors. To say that your processes will fit into a standard set of 20 procedures ignores the reality of parallel functions where product may be handled differently. For instance, your company may use two different warehouse areas for product depending upon whether the product is customer-owned or company-owned, and the two different departments could be performing those warehouse functions differently. You should be particularly wary of an organization or person that accepts the project based upon a single lump sum fee, since they are now racing the clock to provide the ISO manuals to you in as short a time as possible.

The same words of caution apply to consultants. Ensuring that the consultant you hire has helped other companies achieve their ISO certificate and that the consultant understands your industry goes a long way toward being satisfied with that consultant's performance.

Costs for consultants are all over the map. Some use hourly fees of $50.00 per hour on up into the hundreds of dollars per hour. Others provide their services for a set fee based upon a specific task, such as providing in-house training. While cost and budgets are always an important consideration for any company, if you let price dictate what you purchase, you may not get what you want. Consider value rather than cost; quality rather than quantity.

5

Common Pitfalls

There are a number of pitfalls along the path to achieving ISO 9000 certification. Pitfalls, potholes, roadblocks, stumbling blocks, speed bumps (or "sleeping policemen" as speed bumps are so charmingly called in the United Kingdom)—you will most likely get tripped up at some point in the process. What will trip you up will be unique to your particular corporate culture. Nothing seems more amazing than to know that a highly competent set of intelligent, articulate, conscientious people somehow managed to neglect to implement or perform some important aspect of ISO. Hindsight always makes the lapse painfully obvious, but so it is. There's not a company out there that hasn't been caught from behind in some way. Therefore, the trick is to ensure that you stumble as early in the process as possible so that your recovery can be swifter and the impact of your oversight will be minimized.

To this end, numerous internal audits as well as a preassessment audit by the third-party auditor before the first official external audit can help ensure audit success. This chapter highlights some of the most common trouble spots; others are covered in Appendix A.

Zero in on Documentation

The most common oversights revolve around documentation. Documentation is also the area where auditors find the most nonconformances. What do they find?

First, if something is in writing, it should be implemented, and if it is being done, it should be documented. It's extremely easy to document a process, publish it, and then never follow through on the implementation to make sure people are using the new form or following the new steps required by the revised process. This goes back to the common cliché, "easier said than done," but that cliché is more than a little true. It *is* easier to state how you want a process to be performed than it is to check that the process is being performed the way the document specifies. So, be very careful that the processes you change are changed in action as well as on paper.

Ensure that forms are filled out. If there is a box on a form to be checked off or a line requiring a signature or comment, the auditor expects to see those boxes and lines filled in. If the boxes and lines are not necessary, then they should not be included and the form should be redesigned without them. Don't continue using an outdated form just because you are comfortable with it. ISO doesn't care about comfort. Update the form, document the update, keep the new form in a controlled form binder, and ensure that copies of all the old forms have been removed from circulation.

Sometimes errors occur when there is a handoff between departments. Both departments assume the process is written into the other department's processes. "I just forward the form to Shipping," you might be told. Or, "I tell Purchasing when they need to do an inspection; I don't know what forms they use." Or, "That's done before I get it," might be another answer. Unfortunately, each department thinks the other one is performing a task and so it doesn't get done. So be sure to check all loops and handoffs between departments to ensure process closure.

Conversely, many organizations overlook documenting minor "invisible" procedures. These invisible procedures can consist of processes such as auditing the internal audit procedure. After all, you need to confirm that the internal audit procedure is being performed correctly, too. A neat solution to this issue is to have a consultant perform the audit of your internal audit function. Keep an eye out for hidden processes like this which are easily overlooked.

Follow-up Purchasing Issues

Next, purchasing and subcontractor issues are favorite stumbling blocks. The ISO 9000 standard requires that you monitor incoming materials to ensure that those materials are of a consistent quality. This is easiest to do, and most typically done, by formalizing the vendor assessment and review process, and publishing an Approved Vendor List. The process for reviewing vendors, checking product, and determining quality levels that justifies placing certain vendors on the dock-to-stock portion of your list must all be written down and followed. Unfortunately, as time goes by many companies grow lax in checking incoming materials, assessing vendors, or in maintaining the Approved Vendor List.

Anything can happen, even when processes are implemented with the best of intentions. For instance, one company authorized Approved Purchasing Agents in other departments within the company, then was unable to control the purchasing activities of those purchasing agents. This was because incoming product quality was never reported back, nor did the purchasing agents use the Approved Vendor List, nor were vendor assessments conducted on new vendors because the purchasing agents didn't perform them, nor was Purchasing notified when these were necessary. The purchasing agents in this case were trained but not adequately monitored. Whether the function remains centralized or decentralized, the requirements for assessing vendors and incoming purchases as well as who can generate a purchase order must be clearly identified, and the people performing the tasks must be trained and monitored in all facets of the process.

Provide and Track Training

This brings up another sore spot—training. One of the key requirements of ISO is that the person performing a task must know how to do it correctly. There may be certain educational requirements that the person must possess to even be assigned the job, but once assigned, the training that the person receives to ensure that he or she is performing the job according to the documented procedure is of equal if not greater importance to the auditor. For instance, assembly workers may not need any requirements to do their job other than basic reading skills and the ability to follow directions. In the past, the supervisor would explain how a piece of equipment was to be used, walk the trainee through the process, then let him or her have at it. This is not enough for ISO. Now, the supervisor will need to document that he or she has trained the employee, and that the supervisor has double-checked the trainee's work. To prove that the employee was trained and that he or she understood the training well enough to follow the documented procedure the same way every time, it is now common to administer a written and/or a practical exam. Evidence that these were successfully passed is included in the employee's training records. When nonconformances occur in that employee's work, the employee will need to be retrained to ensure that what was being done wrong now gets done right. A probation period may be appropriate. As mentioned in Chap. 4, after ISO has been in place for a while, nonconformances are generally caused by employees, not poor-quality materials. Even if the raw material was substandard, it is the employee's responsibility not to use it and to place the batch or item on hold for resolution.

To ensure that the training program is adequate may mean developing a minimum set of training requirements for a position, stating when retraining needs to occur, and developing a matrix or chart which shows who received training and when. Refresher courses or retraining that occur annually assure the ISO auditor that people performing the task have current knowledge of the process. Such a view-at-a-glance scenario quickly highlights when training takes place or needs to occur. Again, it's important to consider not just the training the employee needed at the start of the job, but the training that will be required to keep that employee current regarding the tasks he or she performs.

Chris Rew, Marketing Manager for Bio-Rad Laboratories in the United Kingdom, recalls that training procedures were in place but not written down. Rew points out that in the early stages of implementing ISO 9000, Bio-Rad discovered that "work instructions could be streamlined by putting greater emphasis on training. We didn't have to modify our training program, just gather the training records. Certificates aren't necessarily required; on-the-job training can be used to fill in any gaps." Keep in mind that although you do not need to present copies of a person's educational diploma to the auditor, you do need to provide objective evidence in some fashion that training was been received and passed.

Resolve Nonconformances

One of the more interesting conundrums of the ISO 9000 quality management system is the triangle of interlocking elements regarding nonconforming product, corrective and preventive action, and statistical techniques. Each part of that triangle relies on the other two parts in order to have any merit. For instance, when nonconforming product is identified, it is relatively simple to isolate the product, perform tests on the rest of the batch to determine the extent of the nonconformance, and get on with the task of making conforming product. But what's to stop your organization from making the same mistake again? Corrective and preventive action are designed to prevent a recurrence of the problem.

Corrective action looks at the nonconformance and provides a quick fix for the interim while seeking to uncover the root cause for a long-term preventive action fix. The time period for resolving a corrective action needs to be stated and needs to be relevant. In other words, most corrective actions should be addressed, applied, and signed off within a few weeks of the occurrence of the nonconformance in order to ensure that the cause of the nonconformance is addressed and the problem doesn't get repeated. For example, if the problem was with the raw material, the incoming inspection process should be closely scrutinized to ensure that substandard raw material doesn't make it past the receiving dock again. Even if the nonconforming raw materials do not adversely affect the quality of the end product, the need to stop the assembly line costs the company money in overtime and delays product delivery. Best not to let it happen, period. And, the Purchasing department may need to reassess the vendor to ensure that the vendor's production practices are also of a level to ensure that substandard raw materials are a fluke rather than a regular occurrence. If the vendor does not have adequate controls over product quality, then the organization may need to consider purchasing from a different vendor. Corrective action, therefore, can have a wide-ranging impact on several areas of the organization. If the way in which the nonconformance is addressed is inadequate, this will show up in repeat nonconformances of the same nature. If the same nonconformance continuously recurs, the third-party auditor will take note and question why the root cause is not being identified and resolved. In turn, the organization could receive a nonconformance during an external audit for not adequately addressing the root cause of those nonconformances.

While ISO 9000 does not mandate that a company implement continuous improvement measures, it does require that the organization implement some level of statistical analysis and tracking techniques. Most companies set goals to either reduce the number of defects, or expedite product build times, or improve the process in some way. This is continuous improvement. How many statistics the company wishes to track is left up to the individual company. However, using statistical techniques can confirm that certain nonconformances are being reduced over time. This provides an incentive to strive for zero defects that can ensure a higher-quality product. Will your customers know the difference? Of course. They will, in turn, have fewer problems using your products for their own processes and will be more inclined to buy from your company again in the future.

Statistical techniques can prove that the corrective action in response to the nonconformance was successful. Without statistical tracking and monitoring of a process, you can never be absolutely certain that defects are being reduced and nonconformances eliminated. Without corrective action, the nonconformance will in all likelihood recur. Of course, it would be ideal to eradicate nonconformances completely—which would make statistical techniques and corrective action unnecessary—but errors *will* occasionally occur. However, as time goes on and your statistics report that you are reducing the number of defects, it may be possible to reduce the number of checks for defects in a process. Many companies have been able to expedite processes simply by finding and eradicating certain types of errors that no longer need to be monitored since they no longer occur. Efforts can then be directed to other areas for improvement. As you can see, each facet of this triangle has a significant impact on the other two elements, and not all of it is negative despite the negative nature of these criteria.

Conduct Thorough Internal Audits

Another area where organizations have difficulty is in ensuring the adequacy of, and follow-up to, internal audits. Part of the problem is that internal audits are typically a new process. But another part of the problem is that resolving the issues which the internal audit uncovers requires the cooperation and response of the department that received any nonconformances during the audit. It is not the fault of the internal auditors if the department under question does not respond and resolve the issue with a corrective action within the agreed upon timeframe. However, because the nonconformance was issued by another department, many times it is pushed aside as a lower priority in the face of the daily demands to perform the job at hand. Unfortunately, this "fire-fighting" mentality is exactly why people are making errors in the first place.

There are a couple of underlying issues that often prevent internal audits from being as effective as possible. First, the issue of internal audit adequacy must be addressed. If your internal auditors do not uncover nonconformances, then they are not examining the process under review in sufficient detail. One quality manager has been overheard to say that he wants to have nonconformances uncovered during every audit because it keeps people on their toes. In fact, your internal auditors should be able to find more nonconformances than an external auditor because people inside the company know more about potential problem areas that are specific to the organization. Given this, internal audits should seek to root out those problem areas and bring them to the light for resolution.

Second, the issue of following up on those problem areas and nonconformances after the internal audit can often be another pitfall. The internal auditor found a nonconformance, it was written up, and the department supervisor specified a corrective action—but that was the last activity to be recorded. One of two things happened. Either the corrective action was performed but the Nonconformance Report was never signed off and returned to the internal auditor, or the corrective action was never implemented. Follow-up is also often lost in the shuffle because performing internal audits is seldom the main job responsibility of the internal auditor. He or she files the nonconformance and gets back to work, expecting the department that received the nonconformance to follow through.

These two types of neglect highlight the fact that there must be some form of responsibility and accountability for internal audit adequacy and follow-up. The ISO Project Leader must have the status and clout to ensure that internal audits and ISO 9000 requirements are not relegated to the back burner. In turn, upper management has the responsibility to set an example by placing ISO 9000 quality issues ahead of other issues such as moving product out the door. This means that sometimes orders will be late. No one in the organization finds this an acceptable answer, but is shipping substandard product on time any more acceptable? The important thing is to eradicate the root cause of nonconformances, not repeat the problem because there isn't enough time to stop.

When implementing your internal audit procedure, you can approach the task with zest, as Bio-Rad's ECS division did. John Goetz, Division Manager for Bio-Rad's ECS division, recalls that the division "had a full-blown audit of our quality system that took two and a half days. It was really tough but we got a lot of good information."

"We found many opportunities for continuous improvement," he emphasizes. "This is good because it looks at the process from a positive viewpoint so that nonconformances and complaints get to the root cause."

Calibrate Equipment

Another area where third-party auditors find many nonconformances is in the calibration of measuring and test equipment. This is such a straightforward issue that it seems odd to find nonconformances here. Of course the occasional piece of equipment will sometimes fall out of calibration ahead of schedule, particularly those that receive heavy use. In the case of heavy use, the calibration frequency could be increased. However, calibration nonconformances do not so much occur because the equipment is out of calibration, but because equipment that should be calibrated is left off the list or because the calibration is insufficient to measure the functionality of the equipment. Calibration is one of the biggest auditor hot buttons. The question the external auditor will ask is "how do you know?" and "what reference standard are you using to ensure that the calibration measurement is adequate?" The auditor wants a record of all the details, whether in or out of spec, and by how much or how little, not just a note stating "in spec."

Take a thorough look around all areas of the company to review what equipment, including software, is being used for any kind of measurement, test, or inspection process. For instance, look for weight scales. Weight scales are usually found in receiving, assembly, packaging, and shipping. Look for measurement tools such as micrometers, which are commonly found in the hardware or MIS departments as well as in research and production. Inspection equipment can be found not only in production, receiving, and shipping, but also in purchasing, quality assurance, and other areas that perform incoming inspection as well as in-process inspection.

Particularly in laboratories associated with large production facilities, such as QA/QC labs, software will be running tests and analysis of chemical batches. In this case, the software needs to be validated to ensure that the measurements it makes are accurate. There is usually a standard routine which the manufacturer provides for testing the accuracy of the software; this needs to be run on a regular basis.

If the equipment is self-testing or self-validating, again the auditor will want to know how you know that the self-testing routine is accurate. Again, the equipment manufacturer should be able to help with an analysis routine that checks the accuracy of the self-test, or the manufacturer can become part of your validation routine by performing the second-level validation check on a regular basis and keeping the associated test records and calibration up to date. Don't forget that when you use an outside organization to perform calibration, that company is a vendor and must be on the Approved Vendor List, and all records must be available on your site, not theirs. So copies must be provided of all the tests, calibrations, and measurements that the calibration vendor performs. And, of course, the calibration vendor will need to update the calibration tags on the equipment as well.

Develop a History Trail

The theory behind a documentation system, particularly a quality management system such as ISO 9000 that relies so much on documentation, is that if it is in writing you have proof—or objective evidence as ISO auditors like to call it—that a task has been defined and performed consistently. Tasks or processes that are defined and performed the same way every time are considered controlled processes. If you can't qualify how a process was done each time, it is considered uncontrolled and, hence, out of control.

In addition to specifying how a task should be performed, recording the process that has been documented by signing off checklists and filling out forms ensures that you also have documented evidence that the process is and was actually done. In turn, this documented evidence provides a history trail which the auditors use to confirm that you are indeed doing what you said you would do.

This history trail, or audit trail as it is also called, is necessary to prove process consistency and completion. The auditor will review a process or function and ask how it is done. He or she will then want to know if it was done that way the last time and ask to see the records that prove it. Those records should be reasonably close at hand. Even if the records aren't kept in the same department, the person completing the form should be aware of where the completed forms are filed. Sometimes everything is maintained electronically and access is as simple as pulling up a file. It's important that the person know where the record is without hesitation and can direct the auditor to the record's location, or show the auditor the record, or tell the auditor where records are sent after they leave the department.

The history trail can be a stumbling block for some organizations for a variety of reasons. Sometimes the forms and records are part of the implementation process and travel with the product from department to department. Departments that perform tasks do not always have copies of the forms. In this case, they need to know where the form is sent at the very least.

Another area where document history trails often fall short is for items such as vendor assessment and training records. Proof of competency in these cases can be "grandfathered" in. This is acceptable initially if the person has been performing the tasks for many years and is obviously competent in that performance, or if the vendor has been supplying acceptable product for an extended period of time. Still, a minimum training review must be conducted to ensure that a record is on file which supports the verbal claim that the person does indeed know it all. The same is true for vendors and subcontractors that have been supplying satisfactory product for some time. Eventually, however, all vendors that your company purchases from will need to be assessed and monitored.

One way to address the history trail issue is to set a date prior to the preassessment audit—usually about three months out—to mark the accumulation of all records associated with ISO 9000 that are to be filled out and maintained according to the ISO criteria that has been documented. The calibration records must be in place. The incoming inspections must be performed and documented. The paperwork that follows an order from one department to another must have all the proper signatures and dates and then be filed in the way your documentation specifies. Your organization can't be audited if there are no records to audit. If there are few or insufficient records, you will receive a major nonconformance and the auditor will stop the audit on the spot.

Review Processes Against the Standard

All of the common pitfalls mentioned in this chapter can be easily avoided. However, the very simplicity of the pitfalls is the reason they were overlooked. It is difficult to understand exactly what more is expected when a system of this nature is overlaid onto existing procedures. The extra documentation and approvals required seem like excess bureaucracy to the people performing the tasks, and they increase the workload of the people assigned the additional ISO-related duties. Because the standard is so generic, and, for instance, just notes that vendors should be assessed but leaves it up to the organization to determine how to do it, many organizations believe that they are already performing tasks such as vendor assessments satisfactorily. However, it is rare that those tasks are being performed consistently and often the current procedure leaves out key criteria such as initialed and dated approvals. A close look at almost any department reveals problem areas that ISO requirements could iron out.

The familiar adage, "Keep it simple," can come in handy. That's exactly how Bio-Rad dealt with the issue in the early stages of their ISO implementation at the ECS division. "ISO showed a lot of areas that needed to be closed; a lot of gaps in our processes," Goetz states. "Processes take time to perfect. We realized we could end up with even more bureaucracy, so we focused on keeping things simplified."

Solutions to Pitfalls

Use the ISO 9000 standard to improve your ISO quality system—perform numerous internal audits. Take advantage of the preassessment audit by the third-party auditor to root out problem areas. Listen closely to the auditor during the preassessment—if the auditor makes a suggestion, follow up on that suggestion because the auditor will look for that follow-up during the first official audit and write a nonconformance if that suggestion isn't implemented.

Ask the obvious questions during your review. Ask yourself: "How do I know that a task was done?" "How do I know that the task was done correctly?" "Where is the objective evidence?"

In order to know what more needs to be done above and beyond the current practice, it is necessary to review each process against the standard. Keep in mind when reviewing the purchasing function that it needs to conform not just to Section 4.6, but to the requirements for documentation, training, records review and maintenance, incoming inspections, etc. Therefore, you must review each process against the entire standard, not just the obviously applicable section.

6

Preparing for and Achieving Registration

Preparing for and achieving registration to the ISO 9000 standard is composed of numerous steps and is certainly not independent of the activities described in previous chapters. But once you've assessed your quality system, implemented the documentation and processes necessary to comply with ISO 9000, and performed your initial internal audits, you're getting very close to bringing the third-party registrar onto your site. This chapter covers the issues surrounding those elements of the registration process, from selecting a registrar to preparing for and surviving the external audit.

Choosing a Registrar

A registrar is a third-party organization that audits and registers your quality system to the ISO 9000 standard. Selecting a registrar can take place as soon as you decide to pursue certification; in fact, you should do so early in the process since the more popular registrars are often booked months in advance.

Interestingly, the relationship between a company and its registrar is an ambiguous one. On the one hand, the company is paying the registrar to perform a service. On the other hand, that service may not result in an answer the company wants to hear. Because registrars are not allowed to offer consulting services (although they may provide training), they should not offer suggestions or direction on how to create or revise your processes to comply with the standard. In addition, some registrars are recognized by an accreditation board and some are not. As a result, choosing a registrar can be a lengthy and time-consuming task.

Because the ISO 9000 standard gained popularity so rapidly, the number of companies offering registration services also grew exponentially. Finally, in 1989 the American Society for Quality Control (ASQC) put a lid on the burgeoning industry by establishing a single registration body that would be responsible for accreditation of the registrars. This organization is the Registrar Accreditation Board (RAB).

It is the RAB's charter to oversee the competency of U.S.-based registrars through a rigorous system of evaluation and accreditation. The registrars that the RAB has accredited include not just U.S.-based organizations, but numerous branches of established European registrars. As of October 1995 (the latest formal figures provided by the RAB), some 30 registrars have been accredited, with 8 pending applicants. Prior to that time, there were approximately 50 "registrars" operating in the United States—so you can understand how easily it would have been to choose one that lacked the appropriate accreditation to ensure the validity of the ISO registrations they were issuing.

Regardless of whether you choose a national or foreign registrar, you should weigh a number of criteria before making your final selection.

Registrar accreditation

It is extremely important that you ask the registrar about the company's accreditation, recognition, and auditor certification. It's almost too painful to contemplate, but in the early days of ISO 9000 registration, some companies discovered too late that their registrar was not accredited or recognized. As a result, their registration wasn't recognized either. After months of hard work, it was back to square one.

There are several accreditation bodies in Europe that accredit European registrars. The two largest and oldest that certify organizations to perform third-party quality system audits are the Raad voorde Certificatie (RvC) in the Netherlands and the National Accreditation Council for Certification Bodies (NACCB) in the United Kingdom. Thanks to the failure of the European Community (EC) nations to create a single pan-European accreditation program, there is no single registrar which can ensure ISO 9000 compliance in Europe, let alone around the world.

The RAB was developed in the United States to establish credibility for U.S.-based registrars. The RAB conducts initial audits of registrars, issues certificates of accreditation, and performs regular follow-up surveillance of registrars. The RAB is formally recognized in the Netherlands, the United Kingdom, Japan, Australia, and New Zealand by a Memorandum of Understanding (MOU), which is an agreement between two third-party organizations for reciprocal recognition of their quality system certificates. The MOUs acknowledge that each accreditation body is the appropriate party to work with on accreditation matters, and define the steps for achieving mutual recognition between those bodies.

As can be seen by the scarcity of European countries in the previous paragraph, there appears to be a reluctance by the Europeans to recognize the United States, despite the fact that the ASQC was on the ISO board. Three years ago authorities predicted that the harmonization of the RAB, RvC, NACCB, etc., under one board by ISO would occur within two to three years. Unfortunately, this has not been the case. However, the wording in the various country adaptations of the ISO 9000 standards follows the original text almost verbatim, and the accredited registrars around the world perform intense audits regardless of their country of origin. What has happened as a result is an unwritten understanding by the customers driving the process to acknowledge that ISO 9000 certification in one country is virtually indistinguishable from ISO 9000 certification in another.

Despite not currently being formally recognized by the EC or most European governments, the RAB is positioning itself for acceptance once a pan-European program does come into being. This should eventually lead to mutual acceptance of registrar accreditations, but there is no predetermined date for this event. As a result, U.S. companies who sell their products in Europe tend to use U.S.-based branches of EC registrars to ensure market penetration.

Background

With this in mind, one of your first questions should be to determine the source of the registrar's accreditation and where that accreditation is recognized. If the registrar is not recognized, don't enlist that company's services. There can be no possible reason to use an unrecognized, nonaccredited registrar, since doing so will only cost your organization money and delay the achievement of an ISO 9000 certificate that is recognized by the appropriate authorities.

Next, ask the registrar if the company has experience in your industry or market. The last thing you want is to be audited by a registrar with no knowledge of your industry processes. Get a list of companies to whom the registrar has issued certificates, including contact names and telephone numbers. Then, take the time to discover what others say about that registrar.

Availability

One reason it is recommended that you contact registrars quite early in the process is to ensure that the one you want is available when you are ready to be audited. It's important to ask how soon a quality system assessment could be performed. Don't be surprised to hear lead times of six to nine months.

In addition, ask how long the registration period lasts. Typically this is from two to four years. A full assessment occurs at the end of the registration period, so the length of the registration has a direct impact on the frequency and severity of the periodic surveillance audits.

Auditor qualifications

To ensure that you get topflight professionals, ask about the qualifications, training, experience, and education of the audit team. Find out what standards or criteria the registrar uses to qualify and certify auditors, and whether those standards or criteria are recognized in Europe. The best auditors have passed a rigorous Lead Assessor course that is recognized by the International Quality Assessors (IQA) group. Probe to determine whether the registrar subcontracts any of its registration activities, particularly whether their auditors are full-time employees. An organization that uses independent auditors may send a different audit team each time, which could cause needlessly long audits.

Costs

With the formalization of the RAB has come a formalization of certain costs associated with ISO 9000, although these costs can and probably will continue to change. While the focus of costs centers on registrar fees, there are actually three areas where companies spend money on ISO registration: internal costs, external costs, and assessment costs. Internal costs include in-house training, awareness, and time spent writing the manual. External costs include external training and consultants. Assessment costs include the application fee (typically $1000), the desk audit fee (typically $1000), a preassessment fee, and the initial assessment fee (varies), as well as the cost for the certificate (typically $700).

Assessment costs depend on your industry, the number of staff at your site, the number of sites involved, and the activities within the scope of the audit. The minimum cost for a small site ranges from $12,000 to $20,000 if the audit is successful on the first attempt. If you use outside consultants, add $20,000 to $40,000 for the first year. The biggest costs are hidden in overhead and indirect costs, which can soar as high as $200,000 depending on how much change to an existing system is required.

Ask how much the registrar charges. The average cost for a registration audit is $1000 to $1500 per auditor per day. The final cost depends on the extent of the audit, which, in turn, depends on the size and complexity of the organization being audited. Obviously, a four-day audit by several auditors costs more than a two-day audit by a smaller audit team. Find out if travel and expense fees are included or extra. Ask whether the costs for surveillance audits are included in the registration fee or constitute an additional charge.

Find out how many surveillance audits will be performed over the life of the registration, and how long each surveillance will last. When asking about costs, ask about the billing rate and whether you will be charged by the workday or by the hour. Ask whether overtime is applicable and if there is a billing rate for travel time. In addition, find out if travel expenses are additional or included in the rate, whether they are billed at a reasonable rate (no first-class hotels please!), and whether the registrar's auditors will be traveling from a location within the United States or from Europe.

Perhaps most important of all, find out which registrar other industry players used and whether your customers or suppliers have suggestions. Industry experience and recommendations are often the best passport to a successful audit.

Top choices

The more than 30 registrars operating in the United States include a who's who listing of well-known companies from Europe and North America. The United Kingdom's British Standards Institute (BSI) is a popular choice because it wrote BS 5750, upon which the ISO 9000 standard is based. Companies who choose BSI do so because it is extremely well known and respected worldwide. In addition to being the originator of the standard, BSI has also developed numerous other safety and kite marks that have very high recognition and use throughout Europe.

Sometimes the registrar choice is a corporatewide dictum; sometimes the registrars are chosen by the individual units because the company is decentralized. Often registrar selection depends on the location of the site being registered. In addition to BSI, other popular choices include the National Standards Authority of Ireland (NSAI) and National Quality Assurance, Ltd. (NQA). All three registrars have locations worldwide. A short list of the better-known registrars is provided in Appendix E; a more extensive list is available from the RAB.

One of the reasons that companies choose NSAI is because the quality system can be certified all the way through by a single registrar. Since it is the only registration body in Ireland, it has the authority to perform not only ISO 9000 audits but also audit to EC directives for product quality. Many registrars cannot legally perform both functions.

When the variables are equal, registrar selection may boil down to the way in which the registrar conducts business. The relationship with a registrar is a long-term involvement, so selecting one with the same operating style as your own, whether relaxed or intense, often ensures the best fit.

Preparing for and Surviving
an External Audit

Whether you plan to do a preassessment audit or just go for a full-blown formal audit depends on your level of preparation. If you've done numerous internal audits and your registrar has visited your facility, you may be able to bypass the preassessment audit. However, the general feeling is that the relatively low cost of a preassessment audit provides added insurance that you are on track—particularly because many areas of the standard are open to interpretation—and prevents the discovery of unexpected major nonconformances during the final audit.

The preassessment typically addresses each clause of the standard at a reduced penetration level. It should be done at least three months before the final assessment so that corrective action can take place. BSI states they've found that 95 percent of the companies pass the final assessment if a preassessment was done. If you decide not to have a preassessment done, then bump up the number of internal audits you do to accommodate for the difference.

Many companies use the preassessment audit as a touchstone and choose to do a preassessment audit early in the process to confirm that they are going in the right direction. This ensures that if the company missed something during implementation, it won't run into difficulty later. Conversely, if the preassessment is too far in advance of the formal audit, the auditor may neglect to mention something on the assumption that you'll get around to it, but which instead gets overlooked. When you perform a preassessment should depend upon your organization. If it is necessary to light a fire under reluctant implementers, by all means do so early. If it makes more sense to use the preassessment to center attention on the upcoming first external audit, do the preassessment nearer to the end of the process.

Three steps for external audit preparation

Preparing for an external audit involves three steps.

First, you need to schedule the audit. This sounds simplistic, but with the heavy demand for ISO 9000 registrations, many registrars have lead times of six months or more. Select a target date and stick with it, or you may have difficulty rescheduling.

Second, submit your top-level quality policy manual to your registrar for a desk audit about a month before the site audit. What often happens is that when you submit your quality policy manual, the registrar asks for a higher level of detail—things that may have been placed in the procedures. The registrar may also want all second level quality procedure manuals before the audits, which gives them the opportunity to do their homework before they visit you.

Third, prepare the guides who will accompany the auditors during the audit, and prepare your employees for auditor questions. Typically, companies use their internal auditors as guides. Guides should be knowledgeable about ISO 9000, about the business, and about the scope of both the audit and the specific procedure. During audits, guides should only interfere if the auditor moves out of the scope.

"Train people on how to behave when the auditor is in the building," advises Hewlett-Packard's ISO Program Manager Will Cowan. "People are inclined to say too much. Auditors keep asking questions to ensure that they've got to the bottom of an issue. The auditor's question style leads people to think that they should document processes that don't need to be documented. Therefore the guides need to know when and how to say that to the auditors."

It's a good idea to hold a kickoff meeting to review procedures and tell people what to expect during the audit. Focus people on where to look things up. If copying the procedures is not allowed at your company (and it shouldn't be because then people are using uncontrolled documents, which is very dangerous since it annoys the auditors excessively), let people know that it's OK not to know the answer to an auditor's question as long as they know where to look it up. Above all, tell them not to guess.

Here's a good checklist to use: First, get everyone prepared to describe what they're doing with confidence. In particular, ensure they know and can quote the one-page quality policy from your top-level quality manual. Tell your employees that they should only answer the question they are asked during the audit, and not to extrapolate or provide more detail. If they do so, they take the chance that they are providing the auditor with information that may lead to a nonconformance. Second, ensure an orderly, organized production floor at all times, particularly in regard to labeling and location, and make sure that all posted signs and instructions are current. Finally, wrap up any corrective actions and document changes before the audit.

Surviving a third-party audit

The bad news first: Despite the prodigious effort involved, everyone is caught unaware on some aspect of the requirements. That's to be expected and the presence of nonconformances does not particularly reflect negatively on any company. After all, you're paying the auditor to find something wrong. With a standard so open to interpretation as ISO 9000, that's exactly what will happen.

To make it even trickier, every auditor has a different perspective and every registrar a different approach. The following advice is based sometimes on cumulative observations, sometimes on an experience that happened to only one organization.

Now for the good news: Virtually everyone survives the audit, fixes their nonconformances, and eventually receives a certificate. But be forewarned that you will have nonconformances too. The path to quality system implementation is not easy.

What happens during audits

What gets inspected during an audit depends on your type of business, your industry, and the scope of your operation. If you are a relatively small site of 100 to 200 staff, you will typically host two to three auditors for two to four days. One of those auditors will be the Lead Assessor in charge of the audit. Be prepared to have the auditors arrive early the day your assessment begins. During the opening meeting, the Lead Assessor explains the purpose and scope of the audit and presents an agenda of the areas the auditors will be covering.

Next, the audit team spends an hour or two reviewing your quality manuals. They then split the various functions among them and take off with their guides to examine the processes within those areas. One auditor may visit the department handling contract review, while another may start with quality assurance.

Essentially, auditors look for consistency between what's written and what's in practice. A good auditor will investigate a point, look at relevant documentation, and check for nonconformances. If the auditor finds any nonconformances, the person being audited at that point will be asked to describe how the procedure is performed. The auditor checks what's written against what's being done. If the findings differ, the auditor will ask for an explanation before officially writing down the nonconformance on a report. You may even be able to fix the nonconformance on the spot. Auditors will reword questions to ensure understanding. The object is not to catch the person off guard but to verify the process.

During an audit, the auditors ask you to describe your operations, where you get your directions, and what do you do when there's a problem—not "What does paragraph 2, section 10 say?" They ask the person being audited to "Describe how this procedure is performed?" and "What paperwork do you fill out?" and "If you don't know, where can you find out?" These are all routine questions that the person performing the work should already know. The person being audited will not be asked about tasks outside his or her scope or job function. If this happens, the guide should step in and explain that such tasks are performed by another person or department.

"We found that they looked closest at quality control reviews for documented evidence and evidence of planning and calibration," recalls Geoff Belton, Quality Manager for Fisons Instruments U.K. "They would ask 'show me' or 'could you just get that document'."

"They also wanted to know how we handled corrective actions,"

Belton continues. "Change control to document software specifications was looked at very closely too. [During implementation] we used a standard change control procedure and a standard review and approval procedures for all areas. In some cases this generated more forms but also eliminated many redundant processes."

The audit team will question everyone. They will ask your top manager about the business plan, what his or her form of communication is, whether there were minutes from the last management quality review meeting, what issues were covered, and if there was evidence of closure on action items. The auditors are also looking for a clear definition and statement of responsibility and delegates.

You'll find that the auditors choose a project or order and audit that project or order from start to finish; they don't wander the halls aimlessly. If during the preassessment the auditors find any nonconformances, they will carefully inspect where those nonconformances occurred. One of the silliest, but still serious, nonconformances is using a pencil to write down dates, approvals, part numbers, etc. That's a no-no. Pencil marks are not permanent. Auditors have the knack for finding the weakest point.

"When the auditor finds nonconformances, don't fight these things," advises Jack Kelley, Quality Manager at Mettler's Analytical Instruments division. "If you do, it can keep auditors preoccupied with minor details. And, you don't want to waste good will with auditors."

The auditors will use a checklist to track the clauses that they review in each area, checking off nonconformances and making observations as they go along. Toward the end of the afternoon, the auditors assemble in a closed meeting to discuss their findings. They'll compare what they've found and look for trends. You will also meet with them at the end of each day for a summary of their daily findings.

When the audit is complete, a final meeting is held to announce the results. Most companies are passed "with qualifications" and given a set period of time to address the nonconformances that were found. If you fail at this point, it will generally be caused by a significant collection of minor nonconformances. The auditors will typically stop the auditing process immediately when a major nonconformance is discovered to return at a future date when the problem is fixed. You have the option, however, to ask them to continue at that time if you wish to uncover other areas of weakness.

Nonconformances

How many companies pass an audit the first time? The number varies depending on your source from as low as 30 percent to as high as 75 percent. It also depends upon your interpretation of "pass."

At the end of the external audit, there will be one of three recommendations:

1. *Unqualified registration.*   Few companies have been registered with no nonconformances.
2. *Qualified registration.*   Usually there's a collection of minor nonconformances which the company has 30 to 90 days (depending on the registrar) to address and provide evidence of corrective action.
3. *Fail.*   Companies usually fail due to a significant collection of minor nonconformances, or one or more major nonconformances.

Some companies have passed with 60 to 70 minor nonconformances, but again it depends on the type and scope of those nonconformances. If too many of the nonconformances are of the same type—and what constitutes too many is up to the individual auditor—then the too-numerous minor nonconformances can become a single major nonconformance.

Many companies are under the mistaken impression that if they fail, they'll have to start all over again. This is not the case. It is difficult to fail completely because the organization is given a set time period to fix any nonconformances. Therefore, the nonconformances found during the external audit only extend the time until the organization is registered to the standard.

"We had been under the impression that if we failed, we'd have to do it all over again, not just where the nonconformances arose," explains Mettler's Kelley. "In fact, we had 90 days to fix any nonconformances that the auditors found."

"If it's a minor nonconformance that can be fixed by correcting the documentation," he adds, "all you need to do is send evidence to the auditor by mail that the nonconformance is fixed. If it's a major nonconformance, the auditor comes back for reinspection of that particular item. It reassured us to know that we couldn't flunk the process."

The best audits happen when you have well-prepared guides, the staff is briefed on what to expect, and the auditors don't find any major nonconformances. Eight to ten minor nonconformances is a good audit, but it's not unusual to see 100 nonconformances if the site has 500 to 1000 employees and the audit is a lengthy one which covers numerous departments and functions.

Of the two types of nonconformances, minor nonconformances cover issues such as not signing forms. Major nonconformances include not implementing a process, not writing down what you've done, or writing "not applicable" on a clause that does apply. Major nonconformances can also be a collection of large numbers of minor incidences in the same area, such as document control. A major nonconformance is generally considered to be a breakdown of one of the elements in the ISO 9000 standard. For instance, if a significant number of calibration errors are discovered during the audit, the organization may receive a major nonconformance but still pass the audit if those calibration errors do not seriously affect the product quality as determined during inspections. Documentation nonconformances, however, are viewed much more seriously since without documentation you have no objective proof one way or the other regarding the accuracy of the processes or the output. A major nonconformance for documentation will probably result in a failed audit.

During the final audit, some companies have found that while there weren't any discrepancies in the third level quality work instructions, there have been weaknesses with new procedures and work instructions because there was no history.

Marq Ransom, Director of Regulatory Affairs for Waters Chromatography, points out that "When DQS [Waters's registrar] did a preaudit, they sent an assessor with an engineering background. But for the final audit they sent three people, one of whom was a chemist. The final audit didn't uncover any discrepancies in the third level procedures. They did find weaknesses with the management review procedure because there was no history. Now we have monthly reviews by our QAG teams. This was the only thing they checked during their follow-up visit."

A number of companies have been caught short because they hadn't implemented some of the new procedures and no mention of getting those procedures implemented was covered during the preassessment audit.

"We got feedback at the end of the first day, so we immediately felt better," states Mettler's Kelley. "In the end, we had a dozen minor nonconformances and one major nonconformance for internal audits. We had a chapter in the manual but we hadn't implemented it. This hadn't come out in the preassessment audit, so we weren't prepared, but that's due to the random nature of audits as to what they find and when. It gave me a new perspective on ISO—that internal as well as external enforcement is an important component of the quality system and that this is what puts teeth into the process." Mettler fixed their nonconformances within the required period and was subsequently registered to ISO 9001.

If something is not examined during the original assessment, it will be looked at the next time during one of the subsequent surveillance audits. Nothing escapes. You'll become very sensitive to how your company handles corrective action and nonconformances after the fact, particularly when it's difficult to get all the required signatures. One way that many companies address minor nonconformances is to fix them on the spot during the assessment. This is a smart idea, since this action cuts down on the amount of paperwork required later to address nonconformances once the auditor has left your site.

What typically happens when there are nonconformances is that you will be asked to perform a corrective action, identify the root cause, and notify the auditor of the result. Depending on the extent and severity of the nonconformance, the auditor may accept the written response, or a reassessment or a partial assessment will be performed at a later date to verify corrective action implementation of deficiencies identified during the initial formal audit.

The good news is that after a while the auditors will have difficulty finding nonconformances. Over time, if your processes are fairly static, your organization will only have minor errors such as occasionally finding an old procedure on a wall. However, if your processes are dynamic, and this is the case for most organizations in today's heated-up, globally competitive marketplace, because you are continually reinventing the way you conduct your business, you are also continually revising your processes, and the documentation will continually be subject to change. What's worked well for organizations with dynamic processes is to develop Temporary Change Notices that cover process revisions, and keep the organization compliant with ISO 9000 between the time the new process is devised and formally documented.

When all else fails

When all else fails, you have a last resort. There's nothing stopping you from disagreeing with auditors and defending your procedures. This is because one of the problems with the standard is that it is not looking for efficiency or effectiveness. Remember that you are the final decision maker on how to run your business, and that's why it comes down to interpretation of the standard. If you don't agree with what the registrar says, you can send their decision to be arbitrated.

Because the standard was written to be generic to all industries, it was written in a general style that leaves much of it open to interpretation. While each organization will implement the standard in a manner appropriate to their organizational style and structure, thus interpreting it in their own way, the third-party auditor will also have his or her own interpretation of how the standard should be implemented.

In a worst-case scenario, the auditor will submit a list of nonconformances that your organization finds unacceptable for one reason or another. You have the right to challenge the auditor. Many organizations have argued for their interpretation, even challenged the auditor's assessment, but that energy is ill-spent. Here's why: Auditors are trained to not allow you to argue your case during the closing meeting. The opportunity to do so was given during the assessment; therefore, in the auditor's eyes, if you couldn't defend the issue at the time it was found, you're wasting your breath and his or her time during the closing meeting. You can challenge the assessment, however, by contacting the auditor's superiors and asking for a review. If you dislike that response, you can take it up with the accreditation board that certified that particular registrar. This is a long process and you may actually get what you want in the end, but what have you accomplished? You've spent significant time, energy, and money on an issue that will always be a thorn in your side. Because of the ill will, you may decide to use another third-party registrar. What if that registrar picks up on the same issues as the first registrar you used? You're right back where you started.

While the intent of the standard is most definitely to implement a quality system that can be proven effective by an external auditor, there are going to be issues, procedures, documentation, and more which you feel are redundant and unnecessary but which the standard requires. If the advice you receive says implement it, and it doesn't negatively affect your product or production processes, it's best to heed that advice.

For instance, the calibration requirement is an irritant to every organization that implements ISO 9000. It's one of the more time-consuming, paper-intensive requirements. As such, many organizations try to exclude certain equipment from the calibration requirement. Thermometers, for example, can be claimed to be used for reference purposes only, and that the actual temperature they measure is irrelevant. If this is the case, why measure at all? Rather than argue this point with the auditor, it's not that much extra effort to check the accuracy of the thermometer on a regular basis. Once this procedure is implemented, it takes a minimal amount of time and keeps the auditor happy. Put your energy into surmounting obstacles, not confronting them.

Eight hot buttons auditors always examine

1. *Quality system.*   Particular attention is paid to the quality policy. It can't just be a piece of paper on the wall; your employees need to be able to state what it is and what it means to them. Don't make it so complicated they can't remember it, or worse, explain it to the auditor.

2. *Management review.*   Make sure that the agenda for management reviews is defined in the procedures, and conduct at least one review before your audit.

3. *Document control.*   Document control is scrutinized more than any other area, particularly whether the documents being used throughout the organization are the latest revision.

4. *Purchasing.*   Implement a formal vendor assessment and monitoring procedure. Develop vendor histories that justify why you buy from that vendor.

5. *Calibration.*   Dispose of all equipment requiring calibration that you no longer use. Don't take a chance on having anything out of calibration. Create a master maintenance schedule and calibrate everything to industry standards.

6. *Internal quality audits.*   Put an internal audit program together and implement it. Cross-train staff to perform audits and hold at least one internal audit before your external auditor arrives.

7. *Corrective action.*   Ensure that corrective actions are conducted in a timely manner, documented, and signed off. It's a good idea to have them reviewed for trends and process improvements during quality meetings and management reviews.

8. *Training.*   Ensure that all your training records are current, up-to-date, and easy to locate. Make sure that employees have completed necessary training, and that the trainers are qualified to teach.

7

Beyond Certification

Once registration to ISO 9000 has been achieved, the system cannot be put aside or relegated to a secondary status. There are a couple of features that enable an ISO 9000 system to keep from being put on the back burner. One of these is the requirement to manage the ISO-created processes which were put in place to support the program, such as management reviews and internal audits. These activities must occur and be documented in order for the ISO registration to remain valid. Indeed, making a successful transition from seeking registration to postregistration operations depends on these processes.

Another reason ISO 9000 activities will continue to have priority is the follow-up surveillance audits. In order to maintain the registration, the organization must undergo repeated third-party audits, generally at the rate of one to four times a year. Each time the auditors return there is always the possibility that you will fail the audit and your registration will be canceled, suspended, or withdrawn.

Follow-up Surveillance Audits

An ISO 9000 certificate is good for two to four years depending on the registrar. The surveillance audits will occur as seldom as once a year or as frequently as once a quarter depending on the results of the initial assessment and the policies of the particular registrar. If the surveillance audit occurs annually, then it will be almost as extensive as the first formal audit. During follow-up surveillance audits, each time an assessment is conducted, certain clauses of the standard are always examined. The remaining clauses are covered during the period in which the ISO certification is valid so that there is a complete system reassessment during that time.

The surveillance audits do not get easier over time. The auditor has the leisure to examine areas in depth that may have been covered lightly during the initial formal assessment. While the ISO 9000 certificate is valid, however, if the registrar continuously assesses your organization during that time period, then the renewal audit which confirms recertification and reregistration to the standard will not be much more extensive than any other.

If you do let the ISO 9000 system lapse and fail to pass an audit, your registrar can cancel, suspend, or withdraw your registration. If they do, find out what their policy is regarding suspended registrations, how you will be notified, and if the registrar publishes the quality system suspension, cancellation, or withdrawal. Such an event should not take you by surprise, because most organizations are quite aware of the status of their quality system.

In the beginning it was believed that all a company needed to do for surveillance audits was to keep the documentation up to date. There's a bit more to it than that. Dynamic companies are constantly developing new products and new processes to go with those products, so there is always something new to audit. This means there will always be changes to the documentation, providing new material for the auditor to review, particularly when your organization is actively performing process improvement.

ISO System Is a Foundation

The goal of many organizations today is to continuously introduce new products into their product pipeline in an effort to stay ahead of their competitors and to discover new or broader markets. Even for companies in mature markets, the cost containment pressures alone may drive the development of new processes when previous departments are discontinued and tasks are outsourced. New products dictate new procedures. Those new procedures need to be incorporated into the ISO 9000 quality system. The development of new products must be conducted with an eye to ISO requirements regarding design control (4.4), while the revision, elimination, or replacement of quality or product-related activities must be accurately documented in order to satisfy documentation as well as other process requirements.

A well-established quality system can assist in the overall business goals by increasing internal productivity and reducing costs associated with inefficiencies. In addition, the ISO 9000 standard can be used as a foundation for implementing other types of quality systems and for meeting more stringent quality goals. Many companies use the ISO standards as a backbone upon which they apply other quality practices, such as continuous improvement.

Many studies have proven the validity of this expectation that ISO 9000 quality systems can deliver results. Most registrars, for instance, conduct research to support the use of ISO 9000 as a business management tool beyond its obvious use as a quality system framework. Lloyd's Register Quality Assurance (LRQA) published just such a survey in 1993 entitled *Setting Standards for Better Business.*[1] Another survey carried out between 1990 and 1992 by Surrey University provides results consistent with the LRQA research, showing that mechanical engineering manufacturing companies certified to ISO 9000 by LRQA significantly outperformed their competitors across standard business ratios, including profit margin, sales per employee, and asset turnover. Entitled *Fitter Finance,*[2] the Surrey University research shows that registered companies report profits more than double the industry averages, which are 2 percent. Smaller firms in particular averaged 6.8 percent, while medium and large companies recorded more than 4 percent. The sales-per-employee figures for large companies were also significantly higher, with a 95 percent increase reported in this area, or an average of £93,500 per employee.

The Department of Defense (DoD) spends more than $1.5 billion annually to support its quality assurance activities. As of October 1996, the DoD determined that commercial quality standards such as ISO 9000 should replace MIL-Q-9858A where it makes sense. The reasoning behind this was largely supported by their findings concerning detection-based versus prevention-based quality strategies. The DoD found that after a company has successfully implemented a basic commercial quality system such as ISO 9000, "It began to eliminate inspections—and the cost associated with them—and significantly reduced the amount of defects in its products. . . . Companies we visited had dramatic reductions in product defects—ranging from 34 to 90 percent—resulting from these techniques."[3]

Another example of savings using statistical process control found by the DoD concerned the quality measurements taken during a process. In one instance, "It discovered that six quality measurements taken in the process were not critical to the product's quality. The elimination of these six measurements saved a total of 36 labor-hours for each product manufactured. Because of this application of statistical process control, rework decreased from 3 to 5 percent to 0.5 percent, the plant experienced a 50 percent decrease in the product's variability, and a 50 percent decrease in cycle time."[4]

The costs versus the benefits of registering to the ISO 9000 standard has been a topic of much interest since the standards were first introduced. A 1993 survey by Deloitte & Touche of Fairfax, Virginia, was conducted in 1993 on 620 companies registered to the standard to examine this issue. The study attempted to determine the relation-

Sales Volume of Companies	Average Annual Savings	Average Cost per Company *
Less than $11 million	$25,000	$62,300
$11 million–$25 million	$77,000	$131,000
$25 million–$50 million	$69,900	$149,700
$50 million–$100 million	$130,000	$188,800
$100 million–$200 million	$195,000	$208,700
$200 million–$500 million	$227,000	$321,700

* Includes one-time internal, external, and registrar costs

Figure 7.1   The costs and benefits of ISO registration.

ship between the cost of implementation as opposed to the average annual savings that resulted from the implementation. The companies surveyed were of all sizes. The survey revealed that although the initial investment is hefty, becoming registered to ISO 9000 can yield significant annual savings as a result of the improved efficiency and quality that is gained (Fig. 7.1).

The case study on Varian Oncology Systems (U.K.) included in this book further highlights the postregistration adventures and triumphs that can be achieved by an organization that takes ISO 9000 to heart. Varian Oncology Systems in Crawley, U.K., was faced with several tough cost-cutting decisions, including the elimination of internal machine shop activities, which would be subsequently outsourced. The machine shop staff reacted with horror. No one wanted to lose his or her job. The result was the presentation of a plan by the shop staff to revise machine shop processes and purchase new computerized equipment. This plan showed how the company could enable new computerized machine shop activities to continue at less than the cost of outsourcing those activities while retaining greater control over product quality. Varian Oncology Systems (U.K.) used the ISO 9000 system to reenergize their processes and significantly reduce the time it took to build product as well as the defect rate when they did so.

Managing significant change that delivers improvements by orders of magnitude, such as Varian Oncology Systems (U.K.) was able to realize, must be controlled. ISO 9000 provides the framework for such control. This is why it is extremely important to ensure that all the management quality reviews and department quality meetings that were implemented for ISO 9000 continue to occur and be given as much if not greater clout than ever.

The Morton International case study, on the other hand, highlights the ISO 9000 implementation adventures of Morton's Electronic Materials division in Tustin, California. The adoption of ISO 9000 was part of a series of events that began with the need for better control over their laboratory data and documentation. For years, the decision to automate the laboratory data kept getting pushed back. Then, a competitor dispute over intellectual property pointed out the difficulty of locating nonelectronic data. A Laboratory Information Management System (LIMS) was subsequently put in place to manage the data more effectively. Before the LIMS was in place, a concurrent decision to implement ISO 9000 was made. The newly available electronic data helped support the Total Quality Management (TQM) and Statistical Process Control (SPC) programs already in place, as well as effectively launch ISO 9000 at the site. The case study highlights the immediate benefits of ISO 9000 for the laboratory as well as the long-term cultural change that sets the corporation up to take advantage of the dynamic global marketplace.

ISO 9000 certification and continuous quality system validation are ongoing processes that can be leveraged to develop better business processes.

Monitoring Processes with Quality Teams

Many companies approach the task of reducing the number of nonconformances by creating quality teams or quality circles. Creating quality teams also highlights the requirements and expectations of the ISO 9000 quality system. Even if your organization has a quality assurance/quality control (QA/QC) department, you will need to involve people outside the quality department. Indeed, each functional area or department should discuss quality issues during their regularly scheduled meetings.

Quality teams can be culled from different departments or be composed of people from the same department. What the teams should seek to accomplish is devising better ways to perform processes, as well as methods for reducing the number of nonconformances.

In essence, these quality teams seek to drive continuous improvements at the company, taking the original goal of quality-driven processes mandated by ISO 9000 into the realm of best industry practices.

What's Ahead for Quality Systems

Beyond continuous improvement and best industry practices, there are industry dynamics that will also drive the shape and extent of quality system activities within the company. Already ISO 9000 has been through one revision in 1994 that required companies registered to the standard prior to that date to modify their documentation to fit the expanded requirements. The impact was greatest on firms registered to ISO 9002 and 9003. Now, all three standards include the 20 elements, regardless of whether those elements apply to the organization. This means that a manufacturing site must now include the design control element (4.4) in their top-level quality policy but state that they do not perform this function. There should be other revisions to the standard in the future that will, in turn, require revisions to the thousands of ISO 9000 quality manuals that exist today.

There are also industry practices and regulations with which to contend. The U.S. Food and Drug Administration (FDA) performs audits of certain manufacturing businesses, such as the pharmaceutical and food-processing industries. These validation audits check that the organization is conducting business according to the FDA's Good Manufacturing Practices (GMPs) to confirm that the product being sold conforms to the product description and specifications, including legal requirements. A recent revision to the GMPs aligned it closely with the ISO 9000 criteria. Further convergence along these lines can be expected in the future.

In addition, because the ISO 9000 standard is so generic, certain industries are taking the standard and developing subsets of the standard for their own industry practices. The best-known example of this is the QS9000 developed by members of the automotive industry, most notably Ford, Chrysler, and General Motors, to define industry specifications for and replace previous supplier quality programs. QS9000 includes the ISO 9000 criteria but goes beyond the scope of ISO 9001 to include requirements that are common to all three manufacturers as well as requirements specific to each of the "Big 3."

TickIT is a supplementary program created by the Department of Trade and Industry (DTI) and the British Computer Society in the United Kingdom to provide a method for registering software development systems based on the ISO 9000-3 standard. The ISO 9000-3 standard provides guidelines for the application of ISO 9001 to development, supply, and maintenance of software. It was originally written as a guidance standard; the TickIT program turns it into a compliance standard. TickIT is particularly applicable to Laboratory Information Management Systems (LIMS).

EN 46000 is another European standard that was recently adopted to address important industry-specific needs. This standard was designed to supplement ISO 9000 and provides additional requirements for medical device manufacturers who market their products in Europe.

Other industry requirements can include legal directives, such as those being formalized industry by industry in the European marketplace to set standards in safety for products from toys to medical devices. These legal requirements do not specify that an organization must have an ISO 9000 quality system, but it is very difficult to meet the requirements of the legal mandates without one, and ISO 9000 is the only internationally recognized quality standard. Regulations such as these give ISO 9000 even more clout.

ISO is not resting on its laurels either. They are actively developing new international standards such as the ISO 14000 environmental management standard that presents criteria for managing environmental issues within the organization. ISO 14000 is similar in structure to the ISO 9000 standard and can be readily incorporated into the same documentation format. Many companies pursue registration to both standards to present an image of an organization that is both quality and environmentally aware.

The next two chapters provide further details about ISO 14000, the GMPs, and other regulations of interest. It's important to remember that ISO 9000 is just the beginning of the type of requirements that it will be necessary to adopt in order to produce and sell product in the global marketplace. It is not uncommon for large multinational organizations to have the headquarters in one country and manufacturing sites scattered around the globe. Indeed, Nestle, one of the oldest and most respected food-processing companies in the world, conducts only 20 percent of its operations in Switzerland, its headquarters country. All other business is conducted where it makes most sense to conduct it, no matter where that is. Conducting business in other countries requires adherence to the business practices of not only those countries, but the countries where those products may be sold, which again may be elsewhere in the world.

Growing awareness on the part of business and the public concerning the source and manufacture of the products they buy and use is beginning to and will continue to have a greater impact on the methods and means by which those products are developed and manufactured. Regardless of the source, the end users should be confident in expecting that the product they purchase has been manufactured to a certain quality standard—the ISO 9000 quality standard.

Notes

1. *Setting Standards for Better Business,* Lloyd's Register Quality Assurance, Croydon, U.K., 1993.
2. *Fitter Finance,* Surrey University, Surrey, U.K., 1993.
3. and 4. *Best Practices: Commercial Quality Assurance Practices Offer Improvements for DoD,* Letter Report, 08/26/96, GAO/NSAID-96-162, General Accounting Office (GAO), 1996.

The second of the international management standards developed by ISO focuses on environmental issues. Known as ISO 14000, the new environmental management standard made its debut in 1996. Unlike ISO 9000, ISO 14000 zeros in on only one aspect of the organization, environmental management activities, and as a result is significantly shorter. Just how a company will apply the standard will depend on the nature of its activities and the conditions under which it operates. In particular, ISO 14000 focuses on controlled and uncontrolled emissions in the atmosphere, discharges to water, land contamination, and disposal of solid and other wastes. Thus, companies that handle, process, or emit environmentally regulated substances will need to address ISO 14000.

On the surface, ISO 14000 appears to be a politically correct environmental management standard whose time has come. It is designed to benefit not just the organizations that implement an environmental management system (EMS) and the regulatory agencies that ensure compliance to environmental specifications, but the world as a whole. The Technical Committee chartered with bringing ISO 14000 to fruition has kept the standard focused on management issues and prevented the inclusion of performance specifications. However, industry sector standards may arise which outflank that goal.

Any organization that is affected by U.S. Environmental Protection Agency (EPA) mandates and regulations will be affected by ISO 14000.

A Global Environmental Standard

The rapid worldwide adoption of ISO 9000 did not go unnoticed. Eager to follow up on the success of this first international standard, in August 1991 the member bodies of the ISO established a Strategic

```
┌──────────────────────────────────────────────────────────────────────┐
│                        ┌──────────────────────────┐                    │
│                        │        ISO 14000         │                    │
│                        │ Environmental Management │                    │
│                        └──────────────────────────┘                    │
│                                                                        │
│         ┌──────────────────┐              ┌──────────────────┐         │
│         │   Environmental  │              │    Life Cycle    │         │
│         │    Management    │              │    Assessment    │         │
│         │      System      │              │                  │         │
│         └──────────────────┘              └──────────────────┘         │
│   ┌─────────────┬─────────────┐     ┌─────────────┬──────────────────┐ │
│   │Environmental│             │     │             │   Environmental  │ │
│   │ Performance │Environmental│     │Environmental│    Aspects in    │ │
│   │ Evaluation  │  Auditing   │     │  Labeling   │ Product Standards│ │
│   └─────────────┴─────────────┘     └─────────────┴──────────────────┘ │
│      ORGANIZATION EVALUATION            PRODUCT EVALUATION              │
└──────────────────────────────────────────────────────────────────────┘
```

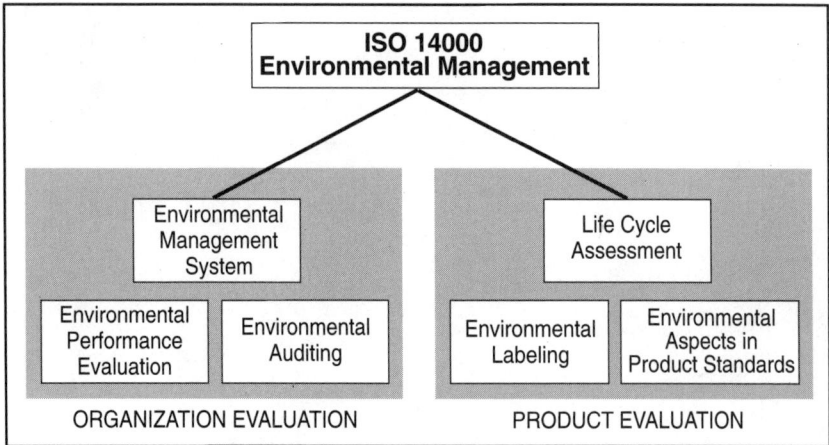

Figure 8.1 The six standards and guides affecting organization and product evaluation issues that make up the ISO 14000 family of environmental management standards.

Advisory Group on the Environment (SAGE) to assess the need for a global environmental initiative. SAGE recommended an overall environmental strategic plan, which resulted in the June 1993 inauguration of ISO Technical Committee (TC) 207 on Environmental Management to develop the ISO 14000 environmental management standard.

TC 207 was chartered to establish standards for environmental management tools and systems, but to exclude any work on performance or the setting of emission or discharge levels. Now, more than 50 nations are actively involved in creating the ISO 14000 series of standards on environmental management. While this figure is comparable to global involvement in the creation of ISO 9000, it's interesting to note that U.S. and Japanese involvement was much more visible in this standard than was the case with the creation of ISO 9000.

In all, there will be five environmental standards. The first was released in late 1996 and included standards for EMS, followed in early 1997 by those on environmental auditing. These will be followed, in turn, throughout 1997 and 1998 by standards on environmental performance evaluation, environmental labeling, life-cycle assessment, and a guide for environmental aspects in product standards (Fig. 8.1).

The Scope of ISO 14000

ISO 14000 is a voluntary standard and similar to ISO 9000 in that it offers management standards and not performance specifications. It is based upon the British Standard BS 7750 and other national models.

Like ISO 9000, which does not call for nor guarantee a quality product, ISO 14000 does not establish required performance levels. In addition, some criteria are virtually identical between the two standards. ISO 14000 emphasizes documentation just as ISO 9000 does, and it has the same type of requirements for an internal auditing function and external audits by a third-party registration body. Currently there are two ISO 14000 standards, ISO 14001 and ISO 14002. Like ISO 9001 and ISO 9002, these two standards reflect slightly different levels of implementation based upon the scope of the organization's practices and involvement in environmental issues (Fig. 8.2).

Companies that handle, process, or emit environmentally regulated substances will need to address ISO 14000. Research and develop-

---

ISO 14001: Environmental Management Systems Standard

1. SCOPE
2. REFERENCES
   2.1 Informative references
3. DEFINITIONS
4. ENVIRONMENTAL MANAGEMENT SYSTEM
   4.0 General
   4.1 Environmental policy
   4.2 Planning
       4.2.1 Environmental aspects
       4.2.2 Legal and other requirements
       4.2.3 Objectives and targets
       4.2.4 Environmental management program(s)
   4.3 Implementation and operation
       4.3.1 Structure and responsibility
       4.3.2 Training, awareness, and competence
       4.3.3 Communication
       4.3.4 Environmental management system documentation
       4.3.5 Document control
       4.3.6 Operational control
       4.3.7 Emergency preparedness and response
   4.4 Checking and corrective action
       4.4.1 Monitoring and measurement
       4.4.2 Nonconformance and corrective and preventive action
       4.4.3 Records
       4.4.4 Environmental management system audit
   4.5 Management review

Figure 8.2 The structure of the ISO 14000 environmental management standard focuses on planning, implementation, checking, and review of the EMS.

ment laboratories which are already in compliance with EPA regulations are prime candidates for implementing the new standard. Such entities will already have processes in place for handling environmentally regulated substances. Formalizing the goals and objectives of these in-house processes should not prove too time consuming or difficult. How the processes should be formalized is not defined. ISO 14000 leaves it to the organization to determine the environmental aspects it can control within the context of its operating environment and regulatory compliance.

If the standard is to be universal and usable by all countries and all sizes of organizations, then specific requirements cannot be included. Joe Cascio, program director for Environmental, Health and Safety Standardization at IBM and chairman of the U.S. Technical Advisory Group (TAG) to TC 207, confirms that ISO 14000 does not detail specific biosphere requirements. He does, however, believe that ISO 14000 should be implemented by organizations that are not required by a legal or regulatory mandate to implement environmental controls. "The standard would apply to any environmentally sensitive or environmentally aware company, whether the organization is required to comply with any regulatory controls or not," he emphasizes.

Legal Implications

Not only is the new standard being promoted as a tool that will enable organizations to demonstrate good environmental practices and remain globally competitive in the environmental technology arena, but equally important, the legal consequences of ignoring environmental issues must be taken into consideration.

According to Jean MacArtor, an environmental lawyer, teacher of environmental law at the University of Delaware, and environmental expert invited to discuss international environmental law with TC 207, "Corporations are being held liable for any negative impact on the environment. Managers, engineers, and other employees responsible for noncompliance with environmental regulations face the risk of criminal prosecution. In fact, the EPA has now adopted a new environmental policy which states that if the company has engaged in criminal behavior, it could be shut down."

Over the past decade, environmental regulations have grown exponentially. Today, the EPA administers some 14 major and hundreds of minor statutes, and more than 10,000 regulations, each of which requires some form of pollution control and abatement. This has affected the bottom line of thousands of companies. The cost of compliance within U.S. industry has been estimated by the EPA to be $122 billion per year.

"One of the ways around enforcement obstacles is to promote prevention and provide incentives for compliance," MacArtor remarks. By controlling the waste streams, the exposure to both U.S. and international regulatory and compliance requirements will be minimized. An effective EMS has the potential to reduce the overall costs of compliance to the company.

In addition to reducing the cost of complying with government regulations, if a mishap occurs that involves a consumer protection issue, ISO 14000 adds another layer of security that the company has done all it could to protect the public. Implementing the standard helps to defend the organization's position that it performed every possible step to ensure against error. The third-party registration confirms this. Thus, implementing ISO 14000 demonstrates that the organization has taken all due precautions.

Implementing ISO 14000

Organizations registered to ISO 9000 should find it relatively easy to implement ISO 14000 because of the relationship between the two standards. The documentation structure is similar and the auditing can often be performed by the same registrar. And, because organizations with ISO 9000 in place are already familiar with the third-party audit process, these organizations are also more likely to add ISO 14000.

It is estimated that companies familiar with ISO 9000 can implement the environmental standard within six months. While ISO 14000 does require a separate manual from ISO 9000, it will fit into the same document control structure. In addition, if organized appropriately, surveillance audits can be done at the same time as ISO 9000 audits.

Everything in ISO 14000 can be dovetailed with ISO 9000. For instance, the mission statement or quality policy can be expanded to include environmental issues. While drafting the EMS documentation, the company needs to look a little further outside its own boundaries by ensuring that it is in compliance with all national and local regulations as well as best industry practices concerning environmental issues.

The extent of the application of the standard will depend on the organization's environmental policy, the nature of its activities, and the conditions under which it operates. To receive ISO 14000 registration, an organization will need to demonstrate to the third-party auditor that it has implemented all the elements of the EMS standard; that it has an effective system for maintaining its compliance to applicable laws and regulations; and that its management practices promote continuous improvement of its systems for environmental protection. As with ISO 9000, the requirement for external audits is what gives strength to the standard as well as ensures that the EMS is credible and functional.

Pros and Cons

The new standard is not without controversy. This is because it focuses on a regulated area of business operations. While some member groups are urging adoption of standards for performance so that there can be a level playing field, this standard has the potential to create trade barriers, especially in Europe.

The goal of avoiding trade barriers may be elusive, since neither this nor any other standard can guarantee a level playing field. In fact, it's likely that some companies will meet resistance when selling into certain markets despite registration to the standard. Cascio explains that requiring a level playing field "means imposing the requirements and systems of advanced industrial economies on the developing world" and has the potential "to make companies in developing nations even less competitive."

On the other hand, individual countries have been creating their own environmental standards for years without much regard to other national standards. This in itself can lead to chaos while providing a serious potential for trade barriers. ISO 14000 seeks to present a common format for addressing environmental issues and is not a performance standard. It was written to minimize the potential impact on trade. If performance specifications can continue to be avoided, the new standard should provide a means for organizations to demonstrate their commitment to environmental policies regardless of location.

"When the ISO 14000 is complete, one international standard should provide harmony in environmental standards worldwide," MacArtor adds. "This harmonization could replace a host of international and regional standards."

Another Jewel in the ISO Crown

While it is highly probable that ISO 14000 will develop and grow in much the same way as ISO 9000, because of the visibility ISO 9000 has generated for quality systems as well as its global acceptance, ISO 14000 should be adopted much more quickly. Cascio expects this standard to be more popular than ISO 9000, "because there are many more people interested in environmental issues than quality ones."

Even so, the drive to implement ISO 14000 will need to come from the marketplace. Competition for customers will need to push this standard in the same way that such competition pushed ISO 9000. And while the push to implement this standard will need to come from the marketplace, the fact that numerous regulatory bodies have been closely involved with this standard since its inception cannot be discounted or brushed aside. Regulatory and governmental bodies around the world will have a vested interest in ensuring the global acceptance of this standard.

Industry pressures may also play a part in the evolution of this standard. While there are no performance specifications in ISO 14000, there are no specifications in ISO 9000 either, and offshoot industry standards such as QS 9000 for the automotive industry have been created that specify certain industry practices. Whether such industry offshoots will also follow for ISO 14000, or whether governments will create such requirements, remains to be seen.

Despite potential market, industry, and government influences, this standard provides an excellent forum for streamlining the environmental management processes within an organization, just as ISO 9000 provided an internationally consistent methodology for quality management. For now, ISO 14000 leaves it to the organization to determine the environmental aspects it can control within the context of its operating environment.

As with ISO 9000, this standard is the most comprehensive effort yet to promote a common international environmental language. As such, ISO 14000 represents today's single best prospect for the development of shared environmental policies and expectations worldwide.

ISO 9000 and the GMPs, GLPs, and GALPs

Quality was an issue long before ISO 9000 arrived on the scene. Quality controls and statistical process techniques have been used to measure product conformity for decades. In the era before ISO 9000 made its debut, W. Edwards Deming pointed the way to continuous improvement that practitioners like Philip Crosby popularized. Most large multinational organizations, particularly those pursuing the prestigious Malcolm Baldrige National Quality Award, tried to achieve continuous improvement and zero defects by implementing some form of quality program to eliminate errors and increase productivity.

Aside from corporate quality goals that also had an eye to improving profits and the bottom line, regulatory criteria have also driven the quality programs at many organizations. Required by the U.S. Food and Drug Administration (FDA), the U.S. Environmental Protection Agency (EPA), and other agencies, these regulations require that certain steps be taken to ensure that the product or process conforms to legal specifications. The most widespread and best known of these are popularly recognized by their acronyms as the GMPs, GLPs, and GALPs.

Good Manufacturing Practices (GMPs)

Good Manufacturing Practices, as GMP stands for, are used by pharmaceutical, medical device, and food manufacturers to define minimum requirements for producing and testing products to ensure that only acceptable product is released to manufacturing and sold to the public.

GMPs are regulations that were originally developed to ensure that medicinal products were produced in a consistent manner and controlled by quality standards appropriate to their intended use. At the turn of the century, prior to the controls that are in place today, the controversies regarding the efficacy and contents of a variety of drugs and medicines caused a great public outcry. Hence, the GMPs were written to ensure that medical products were not adulterated or misbranded.

GMPs define a quality system that manufacturers use as they build quality into their products. Specifically, the regulations define the minimum requirements for the methods to be used in, and the facilities or controls to be used for, the manufacture, processing, packing, or holding of a product to assure that such product meets GMP requirements. In the case of the drug GMPs, this entails meeting certain safety and product identification requirements, as well as ensuring that the drug is of the correct strength and meets specified quality and purity characteristics.

Originally, GMPs were based on the best industry practices. However, as these practices improved and advances in technology were adopted by the industry, the GMPs were revised to follow current practices or current GMPs (cGMPs).

The GMPs covering pharmaceutical drugs were written first and formally introduced in 1893. These drug GMPs are defined in Title 21 of the U.S. CFR (Code of Federal Regulations) under 21 CFR part 210 for drugs general, and 21 CFR part 211 for finished pharmaceuticals. Under these regulations, any drug marketed in the United States must first receive FDA approval and must be manufactured in accordance with the U.S. GMP regulations. Drug GMPs can also apply to veterinary drugs.

The GMPs for drug manufacturing were quickly followed by GMPs for medical devices, known as 21 CFR part 820, and by GMPs for processed food, known as 21 CFR part 110. Under these regulations, domestic or foreign manufacturers of devices and products intended for commercial distribution in the United States were required to have a quality assurance (QA) program. As such, various specifications and controls must be established to ensure that the device or product is safe and effective for the intended use.

Most countries have their own GMPs for drugs and medical devices, although these other GMPs are very similar to those published by the FDA. Most GMPs require proper design and maintenance of equipment and facilities; approved and documented Standard Operating Procedures (SOPs); an independent quality unit; and appropriately trained personnel. There are subtle differences between what the various regulatory bodies require, although the end result is often the same. For instance, the GMPs governing drug manufacture in Canada and the European Community (EC) require internal audits to be conducted. Internal audits are not a requirement of the FDA GMPs. Instead, the FDA GMPs require annual product quality reviews to be conducted. These reviews are to look for trends to determine root causes of nonconformances. Such an examination is not required by the Canadian GMPs.

Whether a drug or a device, the FDA monitors problem data and inspects the manufacturer's operations and product records to determine compliance with the QA program requirements in the regulations.

Current GMPs

The impact of ISO 9000 has caused government agencies around the world to update their regulations to take advantage of growing market acceptance of this international quality standard as well as to strengthen their requirements for compliance. A case in point is the FDA's 1996 update of the medical device GMPs, which greatly expanded the scope of the previous GMP regulation.

Regular revisions to the GMPs are necessary because the changes in industry practices and particularly the adoption of new technology creates situations not anticipated when the original document was drafted. Indeed, the FDA referred heavily to ISO 9000 when revising the GMPs for medical devices because ISO 9000 has evolved into a best industry practice.

The requirements of the original GMPs covered the methods, facilities, and controls used in manufacturing, packing, storing, and installation of medical devices. This included QA programs and organization, buildings, equipment, components, production and process controls, packaging and labeling control, distribution and installation, device evaluation, and records. A QA program meeting the GMP requirements ensured that the finished device met specifications by reducing manufacturing process variation that could lower quality.

The current GMPs strengthen the quality assurance aspects of the medical device regulation by adding controls for design, purchasing, and service, as well as more production and process controls. The requirements for inspection and test status, nonconforming product, corrective action, and statistical techniques have been clarified. As a result, these changes significantly impact the scope of the original requirement.

This latest version of the GMPs closely parallels many elements of the ISO 9000 series standards. This works to the benefit of organizations who achieve ISO 9000 certification. That's because if your organization is in compliance with the proposed GMPs, it will probably be in compliance with ISO 9000, although not necessarily the other way around.

This is because industry interacts with the FDA in two ways: through a product review process and through manufacturing compliance. The product review process is governed by regulatory affairs (including regulations such as the GMPs) and specifies whether the organization can market a product based on information found in that organization's 501K or Pre-Market Approval (PMA). Manufacturing compliance, on the other hand, deals with quality assurance issues, whereby enforcement activities are oriented around the organization's quality system. Thus, while the product review process is subject to design safety and efficacy concerns, manufacturing compliance falls under both mandatory GMP compliance and voluntary ISO 9000 compliance.

Even so, companies already certified to ISO 9000 will have an edge when adopting GMP requirements, as is readily apparent when reviewing the current GMPs.

Other GMP Revisions

The FDA is in the process of revising the GMP regulations for drugs and finished pharmaceuticals. Apparently many organizations were having difficulty providing adequate proof of validation to the FDA auditors, nor were those companies examining or validating their processes in an appropriate manner. As a result, revisions are being made to certain laboratory control and cross-contamination requirements as well as to testing procedures. These amendments clarify manufacturing, quality control, and documentation requirements by revising the requirements for process validation, methods validation, and testing, among others. By describing the steps that should be performed in greater detail, the amendments seek to boost the integrity of the drug manufacturing process and the safety of drug products.

Because process validation monitors and evaluates process performance, it is playing a major role in the cGMP revision. The documented evidence that process validation provides confirms that the manufacturing process is consistent, thereby helping to ensure drug product quality.

The FDA's concern regarding the validity of processes in today's manufacturing environment is caused by the extremely complex nature of these processes—a complexity that will only increase. It was found that unexpected variables in the manufacturing process could greatly alter existing product parameters. For instance, even a minor change, such as to the order in which ingredients were added or to the physical characteristics of an ingredient, could alter the bioavailability of the drug product. Such changes can be difficult to unearth because they are often not tested consistently. The dissolution pattern determined by a research laboratory is a case in point. It might not be the same dissolution criteria being validated during manufacturing. Or, packaging can change the product characteristics. Certain packaging materials have the ability to change the chemical properties of a drug. But once a drug is packaged, the drug is not typically retested before distribution, so the change is not detected. As a result, the purported product is not the one being delivered to the customer. In order to circumvent such situations, the FDA has placed greater emphasis on revalidation after any change in process or product characteristics or control procedures. Companies can address this requirement by ensuring that when a change is proposed, the effect that it is expected to have on the product is also indicated.

The proposed amendments also affect method validation. Method validation is the documented evaluation of an analytical method. Successful method validation provides assurance that the method will consistently yield results that are accurate within established specifications.

Methods validation ensures that the method selected is scientifically sound and that it serves its intended analytical purpose. Accurate methods validation is central to ensuring the reliability of all evidence that supports a product's identity, strength, quality, and purity as defined and required by the GMPs. For test results to be useful, significant, and reliable, the methods used to analyze the data in such test results must also be validated. In other words, an organization must establish that the analytical methods it uses to assess or evaluate a manufacturing process accurately measure variables affecting process control.

Because of methods validation, testing is another area under scrutiny. The proposed rule, therefore, amends procedures for the testing of components, calculation of yield, and blend testing. It also provides more detailed procedures for dealing with out-of-specification results.

To further ensure that validation procedures are current, this proposed rule requires the provision of an independent quality control unit which is responsible for reviewing changes in product, process, equipment, or personnel. This quality control or quality assurance team is also responsible for determining if and when revalidation is required. No doubt revisions to the GMPs for processed foods will soon follow suit.

When the FDA started revising the GMPs in 1993, they solicited input from organizations around the world and from companies engaged in all types of manufacturing. When the previous version of the GMPs is directly compared to the current version, it is obvious that this latest version of the GMPs is better written and more in tune with global quality initiatives. Looked at dispassionately, the proposed GMPs are a good attempt to harmonize with European standards both in language and style—an attribute that will eventually make compliance with any standard or regulation easier to accomplish.

GLPs and GALPs

In addition to the programs put forth by the FDA, the EPA provides regulations that have an impact on the analytical industry in general and the analytical laboratory in particular. The Good Laboratory Practices (GLPs) and Good Automated Laboratory Practices (GALPs) are the two foremost documents among them.

The GLPs present standards of practice which are regulations that govern the management and conduct of most nonclinical laboratory studies submitted to the EPA's Office of Toxic Substances and its Office of Pesticide Programs. These regulations are found under 40 CFR part 160, Federal Register, Vol. 54, No. 158, and were published on August 17, 1989.

The GLPs affect, and are affected by, other regulations. For instance, laboratories that submit studies in support of the registration of pesticides under the Federal Insecticide, Fungicide, and Rodenticide Act (FIFRA) are subject to GLP standards. And laboratories that submit studies required by the test rules and negotiated testing agreements section of the Toxic Substances Control Act (TSCA) are subject to the GLP regulations in 40 CFR part 792.

The GALPs, or 2185—Good Automated Laboratory Practices, provide principles and guidance to regulations, and assist in ensuring data integrity during automated laboratory operations. The GALPs were originally issued in December 1990 by the Office of Information Resources Management (OIRM), an arm of the EPA. Because the GALPs describe benchmarks for assuring the reliability of laboratory data, they are designed to be adopted by laboratories that use or are planning to use automated data collection and management systems, most particularly Laboratory Information Management Systems (LIMS). The GALPs are closely aligned with the GLPs and include many of the GLP requirements.

As industry requirements for higher sample throughput and productivity have grown, fewer laboratory operations are being conducted manually. Because much of the laboratory data being submitted to EPA are now acquired, transferred, processed, managed, or in other ways manipulated by a LIMS, a regulation that addressed this electronic function needed to be developed to ensure data consistency and integrity. Hence, the GALPs were developed to ensure that all LIMS data used by the EPA are reliable and credible.

GALPs address laboratory management, personnel, quality assurance, LIMS raw data, software security, hardware, testing, records retention, and facilities. There are some similarities in these requirements to those embodied in the GMPs, particularly the requirements concerning personnel and facilities. However, the GALPs go beyond the GMPs by venturing into issues surrounding hardware, software, and electronic data management.

ISO 9000 only once mentions, but does not require, that an organization should consider regulatory issues such as these. Nor do these regulations require an ISO 9000 quality system. However, all these systems pursue the same end, that is, to validate the integrity of the process and the product. In the case of the GALPs, the product is the laboratory data. GALPs look for an internal quality assurance unit to perform validation of the system, but this unit does not need to validate the system within a larger quality management system such as ISO 9000. ISO 9000, of course, requires internal audits, but does not specify the existence of an independent quality assurance unit, only that the persons performing the audit be independent of the function being audited. On the other hand, most organizations with an ISO 9000 system in place recognize the importance of a quality program and have an independent quality department. It would undoubtedly make the task of validation easier, and certainly the task of convincing the EPA auditors that the data are valid, if a quality management system such as ISO 9000 were in place.

GxPs and ISO 9000

"In principal, the GxPs and ISO 9000 standards may not seem very different," claims Siri Segalstad of Segalstad Consulting in Oslo, Norway. "They all describe virtually the same quality systems, but their scopes differ considerably. The main difference is the legal aspects of the two systems. The GxPs are ethically motivated to ensure the patient's safety. They are legal regulations that are enforced by official authorities, with sanctions and penalties applied if the regulations are not followed. The ISO 9000 standards are motivated by commercial and marketing reasons. They are not legally binding, but are contractual agreements between the vendor and the purchaser that are privately enforced."

In the real world, both ISO 9000 and the GxPs have a place and meet very specific, different goals. They do tend to complement each other, as the implementation of one can support the implementation of another, and reduce the overall tasks required than if each were implemented independently of the other.

The FDA, in particular, has been keeping close tabs on the changes in the world marketplace, and incorporating those changes into revisions of the standards and regulations it controls.

One requirement the government agencies are keeping an eye on is the use of the CE Mark throughout Europe. The CE Mark can be applied to products that have been developed and manufactured to agreed specifications. It is being used to reduce the barriers between European countries by alleviating the requirement for individual country approval. While applying for a CE Mark does not require ISO 9001 or 9002, having an ISO 9000 system in place can make it easier to prove that the product being CE Marked has been developed and manufactured in an appropriate quality environment.

Because the proposed GMPs are similar to equivalent European standards, U.S. organizations that apply for a CE Mark will be able to continue selling into those European markets that require one. Conversely, if the U.S. organization purchases CE Marked components manufactured outside the United States, then the quality of those purchased components is also ensured. Indeed, it is not getting any easier for a European company to sell a nonquality product into the United States than it is for the United States to sell such products into Europe. With the adoption of the latest version of the current GMPs for medical devices and the upcoming revision to the GMPs for drugs and finished pharmaceuticals, foreign manufacturers must now allow FDA inspections in order to import to the United States.

In regard to implementing a quality system that addresses these myriad product quality requirements, Segalstad offers some advice. "Does the whole process sound rather time-consuming and difficult?" she asks. "It is a lot of documentation to create, but in the end the organization ends up being less dependent on one person or a few persons if everything is written down. The process of writing these documents also makes you think about what you really want to achieve, and how, with your system. Your log books will help you find what you have done and when, and whether any errors have occurred before. The reason for validating the system is *not* that FDA or their counterparts in other countries shall be satisfied. It is a good tool for your organization, and you will feel more confident that the system performs the way it should all the time."

On the Path to TQM

Varian Oncology Systems (U.K.)

One of the biggest complaints about ISO 9000 is that it doesn't address business issues such as cost containment and productivity enhancement. The standard does not provide any direction for how to use the quality management system to drive ways to improve the business. Indeed, it doesn't even require the organization to produce a good product, just a consistent one. However, just because ISO 9000 doesn't define how to improve the business doesn't mean it can't be used to do so. That's exactly what Varian Oncology Systems (Crawley, U.K.) did. They used ISO 9001 to achieve operational excellence while reducing operating costs.

When Varian Oncology Systems was registered to ISO 9001 in 1991, ISO was seen as one of the elements to be folded into the overall corporate charter to implement Total Quality Management (TQM) practices throughout the company.

TQM and Indicators of Excellence

Varian's corporate headquarters in the United States (Varian Corporate) uses "Indicators of Excellence" to evaluate its analytical instrument, semiconductor, and oncology business units two times a year. These indicators use material from the Malcolm Baldrige National Quality Award (MBNQA) to measure how well the business unit is performing. The MBNQA covers a much broader set of criteria than ISO 9000; in particular, it addresses business results.

"One of the downsides to ISO 9000 is that empowerment, process ownership, and other peripheral aspects are not covered by ISO

9001," explains Adrian Orchard, Varian Oncology System's Quality Assurance Manager. "It's unfortunate that many Quality Consultants have misled management into believing they could cut costs and increase productivity with quality measures. In general, this hasn't happened. While you can interpret ISO 9000 however you wish, there needs to be more."

Orchard is not only attuned to the various standards and directives, but active in their development; he sits on TC 210, the international standards committee in charge of quality standards for medical devices. "Right now ISO 9000 doesn't require you to address costs," he points out, "but operational excellence means cost of quality (COQ), defect rates, and other performance indicators, including time to market, must be addressed."

In order to meet the goals of operational excellence set by headquarters, Varian Oncology Systems used the concept of ISO 9000 as a framework to drive operational measures necessary to achieve cost containment and productivity improvements by orders of magnitude.

Initially, Varian Oncology Systems had a smattering of written procedures for manufacturing in place to address the Good Manufacturing Practices (GMPs) required by the U.S. FDA. While documenting the rest of the processes, "we used ISO to rewrite responsibilities," recalls John Peel, Varian Oncology System's Division Manager. "So we broke one of the golden rules that says just write down what you do, because we viewed GMP as a dinosaur from the past due to its formal 'you will do this' approach that was done for defense issues."

"With TQM," Peel continues, "dramatic changes can come about when you benchmark systems. We took this opportunity to change our old centralized QA system and decentralize quality."

First, Varian Oncology Systems redefined quality by functional area. Now the Manufacturing Manager, not the Quality Manager, is responsible for the manufacturing procedures and work instructions.

Then, Varian Oncology Systems focused on teams. "We developed self-directed, cross-functional work teams," Orchard states. "ISO 9000 doesn't say to do this, nor does the standard expect it. ISO 9000 just wants you to maintain the standard you've set. It doesn't encourage you to benchmark or strive for best industry practices."

"Benchmarking helps keep you on track," Peel emphasizes. "We looked at the customer/supplier interface. We looked at other industries for insight. In particular we looked at American Express and Federal Express."

Suppliers were told to become certified to the ISO 9000 standard. "We took our original 400 suppliers down to 100, and most of those are now certified," he adds. "But the process is interactive. Free of charge, we're training the remaining suppliers on what to do. We've tried to shoot for total supplier certification and come pretty close."

As time went on, the roles of the manufacturing staff changed. This became necessary when corporate budget cuts dictated drastic changes. In order for Varian Oncology Systems to remain within budget, manufacturing costs had to be drastically reduced. Interestingly, the manufacturing staff rather than management decided how this was done.

From Following Instructions to Creating Them

"This is the shop floor running the company," Peel points out. "We've become a team-based facility that uses facilitators rather than supervisors." Now, management asks for staff input every time a change, such as implementing new machinery, is considered. "For example, we considered eliminating some jobs and outsourcing part of the manufacturing process," he recalls. "The staff showed us how investing in new machinery could make them more efficient and actually cost less than outsourcing. Now, they're constantly on the look out to improve their processes to the benefit of the bottom line."

According to Steve Pullen, formerly Varian Oncology System's Shop Floor Supervisor and now the Machine Shop Facilitator, "we didn't even know the shop was under threat." When Peel told the manufacturing staff that the machine shop function might be outsourced, and they protested, he told them that the operation could continue in-house only if it could be cost-justified.

Pushing the responsibility for the decision down to the lowest levels meant that the shop floor employees had to take an active role in order to save their jobs. Good jobs are difficult to come by. No one wanted to be out of work.

"We took the problem down to the local pub and discussed it over warm beer and a sandwich," Pullen recalls. "We got together as a team, looked at what outside suppliers were doing, and put together a cost justification," he adds. "We'd never done this before. That cost justification went back and forth between John and us. He pointed out weaknesses and asked for rewrites. We learned all about payback period, overhead, training—you name it. It was an education for all of us. John let us go to other companies to see how they used computerized CNC machining—something that was unheard of seven years ago in Britain. Finally, John approved the purchase and also put in place the JIT [Just-in-Time] system we use today. The key was to bring down inventory levels and then guarantee JIT delivery and reduction in costs."

"We got trained to do programming," he continues. "The boys, five of them, had to learn how to program machines—something none of them had done before. John makes us see the whole process. We even spent a day at a hospital watching the radiotherapy equipment we make being used."

Now there are weekly and daily team meetings on the shop floor that discuss different aspects of the manufacturing process. While one person coordinates the meeting, he is not a team leader. There are no team leaders. Peel eliminated supervisors so that the teams would have to become self-managed and make their decisions as a team.

"The culture shock was tough," Pullen admits. "Now, however, the team has the authority. Everyone has more responsibility and most of them like it. Slowly but surely this is changing the way the company runs."

"ISO 9000 requires a lot of change, from learning about procedures to making the changes," Pullen explains ruefully. "Now, procedures are written by the people who do them. Now, everyone knows exactly what the machine does and what it takes to fix it."

Every team now has its own budget and looks after it carefully. It can be spent on individual purchases or the team can collectively buy a new machine if it wants to.

"Initially, I was very nervous about the outcome," comments Bob Georgeson, Varian Oncology Systems Shop Floor Fitter and former Storekeeper. "When JIT was implemented, we worked ourselves out of a job because there was no more inventory. We kicked up a stink, and were told we had to be retrained. We came up with a proposal and three of us got Fitter training and passed vocational training, two of us are on the shop floor as team coordinators, and two are still in stores processing returns."

Georgeson enjoys the new way of operating. "This is one of the best systems. We see problems first hand and we can fix them. Before we couldn't; now we can. Previously we'd work our way around them. Now, when people find a problem with a part or piece, we can work together to fix it. No more second hand."

"I personally like the change," he continues. "Now we're not so tied down. We can make decisions without having to ask permission the whole time. I see it as a way up in terms of development."

Not only is it a way up in terms of development, but the new system has had the desired effect on operations. Peel points out that productivity has been increased by a factor of ten. "We used to build 18 Ximatrons a year; we can now build 80. There are two reasons we are able to do this. First, we gave suppliers more to do in terms of coordinating an assembly rather than merely supplying components. But most of the productivity increase is due to the men on the shop floor."

Productivity Enhancements
Follow Cost Containment

Aside from the cost-containment objectives they were able to achieve, Varian Oncology Systems has been able to realize some spectacular improvements in productivity as a result of the process reengineering. Again, they used elements of ISO 9000 to frame, manage, and monitor the change. Statistical techniques were used to track and calculate the percent improvements in production and the reduction in discrepancies.

With the change in how the machine shop operations were performed came a corresponding change in the staff roles and responsibilities. Understanding the business case and empowering staff to solve and fix problems has given staff ownership of the processes. With Peel outlining production goals, several spectacular improvements have resulted.

For instance, in 1987, Varian Oncology Systems built nine Ximatron simulators simultaneously over a period of 34 days. By 1994, they were building two to three units simultaneously, but finishing the job in 7.5 days. Not only was product being delivered to the customer faster, but productivity was increased (Fig. C1.1). In addition, Ximatron discrepancies, or nonconformances, per unit at final

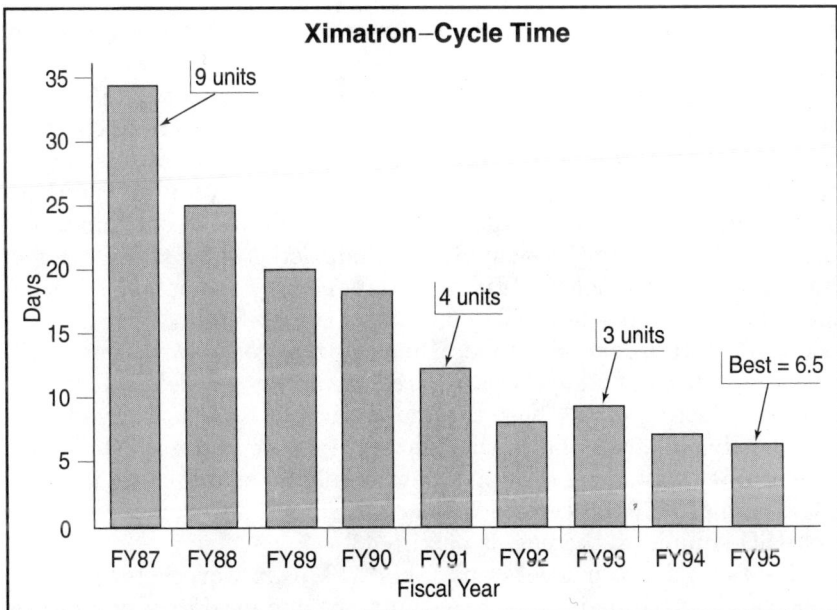

Figure C1.1 Ximatron cycle time production increases.

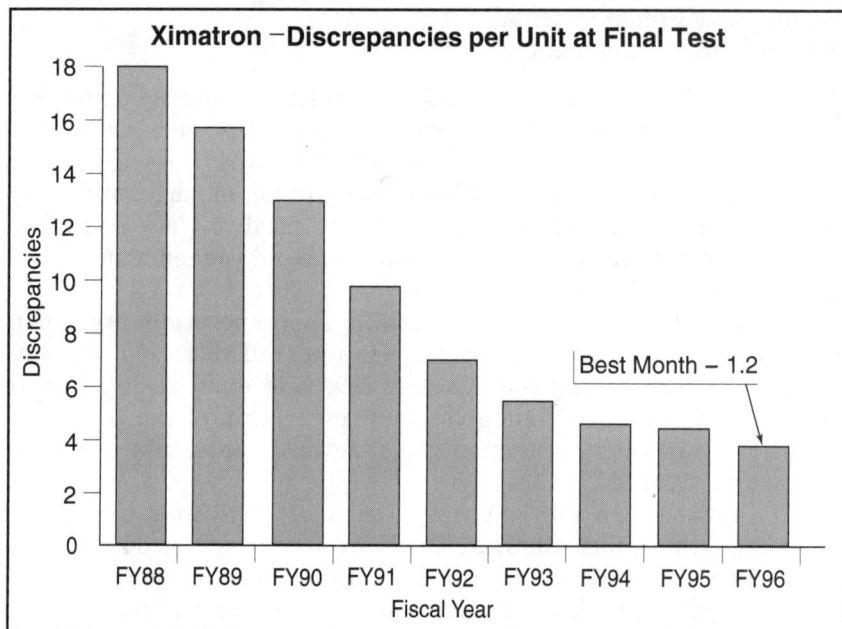

Figure C1.2   The reduction in Ximatron discrepancies per unit at final test.

test were reduced from 18 to 4.2 between 1988 and 1994 (Fig. C1.2) or by 77 percent.

ETR couch production improvements are the most outstanding. Production volume increased while cycle time decreased. Cycle times have gone from 18 days to just 2 (Fig. C1.3), an improvement of 89 percent, while the number of couches produced per year has increased 79 percent. It gets better. With a defect rate of 0.03, or one nonconformance in 30, Varian Oncology Systems' statistics show how close they have come to zero defects (Fig. C1.4). Indeed, by the end of 1996, the site would go on to achieve five months of zero defects!

Supplier discrepancies have also been reduced. "We've got to the point now with our suppliers where we cannot get below 1 percent defects," Peel reports "Those that do get below 1 percent have implemented ISO 9000 similar to this place; they have taken ISO to heart." (Fig. C1.5). "Now, if we take on a new supplier, we absolutely insist on their being ISO 9000 certified. Although sometimes we have to use a non-ISO supplier if the part is patented."

There's no doubt in Peel's mind, nor at Varian Corporate, that ISO 9000 can be used to drive cost containment and productivity improvement measures by significant orders of magnitude.

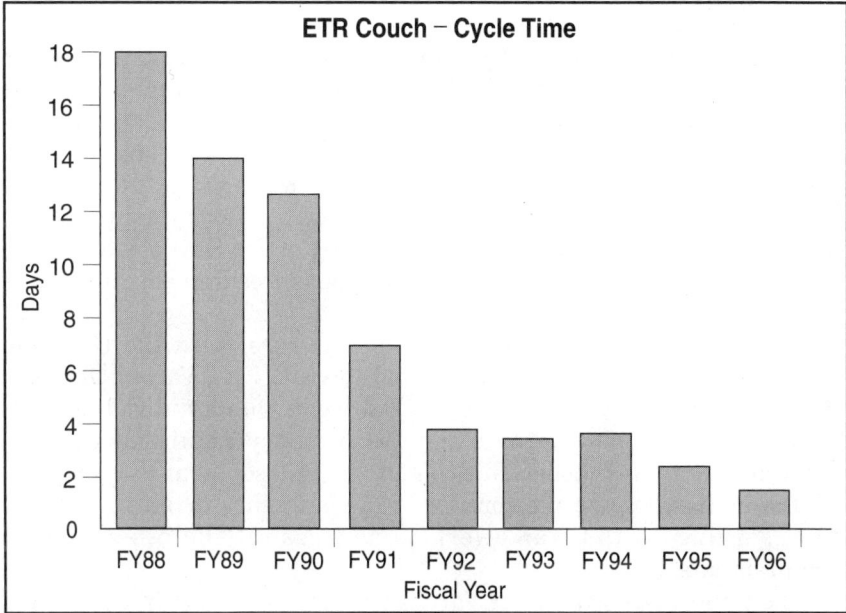

Figure C1.3 Turnaround increases in ETR Couch cycle time.

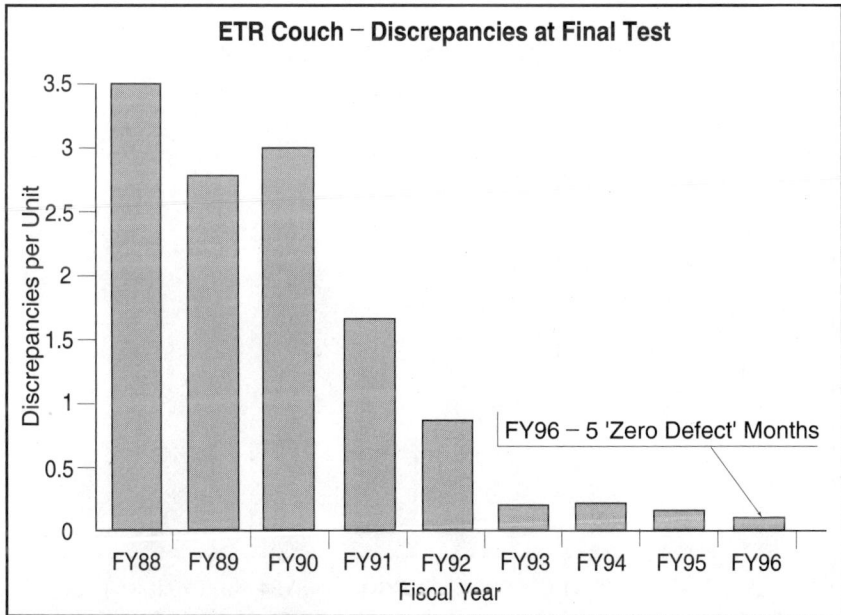

Figure C1.4 The reduction in ETR Couch discrepancies per unit at final test shows the site is closing in on zero defects.

A New Philosophy

The path Varian Oncology Systems followed in 1990 to pursue TQM used a model in which the FDA GMPs were at the core of business practices. The U.K. GMPs were overlaid onto this model, and ISO 9001 was overlaid on top of the GMP requirements. The Malcolm Baldrige National Quality Award (MBNQA) was then overlaid on top of these three elements to address broader business quality issues. The reason for this approach was the legacy of the GMPs, which were the driving force for the original quality measures implemented many years previously (Fig. C1.6).

The 1995 model places ISO 9001 at the core, with the U.K. and FDA GMPs on top of that. And beyond the MBNQA, Varian Oncology Systems is also now addressing the European Quality Model (EFQM) criteria. As is apparent, over time as industry requirements have evolved, so has this model. But so has the philosophy at the company. ISO 9001 has become the core for Varian's quality management system, serving as the framework to address all the other quality requirements.

This philosophy will assist Varian Oncology Systems in addressing other regulatory and legislative quality requirements. For instance, there is a move now in the medical device industry to harmonize on a

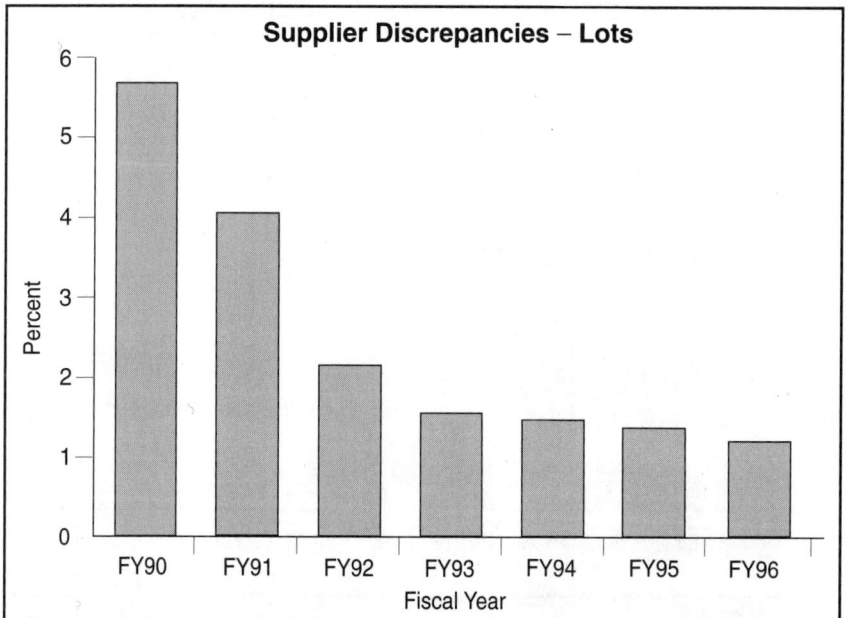

Figure C1.5 The reduction in supplier discrepancies per lot.

1984—Exhortation to action

1985—Perception of crisis forced business paradigm rethinking

1986—Supplier improvement to reduce defects

1987—Corrective action teams to solve big problems

1988—Value-managed relationships to partner with suppliers

1989—Demand flow/process reengineering grew out of JIT

1990—Process ownership

1991—ISO 9001/Baldrige doctrine for customer-focused process management

1991—Continuous process improvement teams

1991—Policy deployment/critical success factors

1992—Process focus rather than department focus

1993—Employee empowerment

1994—Benchmarking

Figure C1.6   The evolution of continuous improvement at Varian Oncology Systems (U.K.).

worldwide standard. TC 210, the technical committee for quality standards for medical devices, has drafted ISO 13485, which was published in 1996. ISO 13485 is a supplement to ISO 9001 and specifies certain quality criteria that medical devices must meet. The quality framework that Varian Oncology Systems developed enabled them to slot the new requirement into their existing system with minimum disruption.

Varian Oncology Systems has also adopted ISO 14000, the environmental management standard, as part of their environmental program. This standard is similar to ISO 9000, and also slots into the existing quality management system.

In addition, the company is actively pursuing the European Quality Award (EQA), the European equivalent to the MBNQA. Varian Oncology Systems has already been a finalist for the EQA.

Thanks to the benefits the company has been able to derive from implementing the ISO 9001 quality system, and using it as a framework to drive not only product improvements but business improvements, they are one of the few organizations that have leveraged ISO 9000 from every angle. "ISO will be here for the long-term," Orchard observes, "But there's more to life than ISO 9001. If standards bodies can devise a way to judge TQM, that will be a big step forward."

Using Quality to Reinvent Lab Operations

Morton International

How do you reinvent the business when your organization is more than 100 years old? A successful operation doesn't get that way overnight, and when you've been a viable market player for more than a hundred years, change which might jeopardize that status isn't implemented lightly. But despite Morton International's precise document control and streamlined processes, there was no question that their manual data handling system had a finite capability and that quality data had to have greater breadth, detail, and accessibility. The question was how and when to implement the change. It took a competitor dispute to generate the first shock wave.

From Salt to Sodium Borohydride

Since 1848, when Morton International of Chicago, Illinois, first started mining and selling salt, the company has grown into a diversified organization. Their venture into specialty chemicals began before the Second World War, but gained strength with the discovery in 1926 of synthetic rubber—the precursor of liquid polysulfide products. Over the years, Morton has branched out into adhesive and specialty polymers, coatings, specialty chemical products, and electronic materials with products such as flexible packaging adhesives and coatings, chemical vapor disposition materials, polysulfide and latex sealants, and sodium borohydride reducing agents.

"We make products that range from the photoresist chemicals used to make Nintendo circuit boards to the circuit boards that go into satellites," explains Phil Lofty, Scientist for Morton's Electronic Materials division in Tustin, California. "Quality is very important. Boards need to be able to pass criteria such as temperature tests from $-40°C$ to $+150°C$ every four seconds for 120 continuous cycles. And when we run a current through the board, we expect it to work every time."

Electronic Materials produces primary imaging photoresists, photoimageable solder masks, diazo phototools, waste treatment processes, auxiliary chemicals, and process equipment. "We're more than half the world's market for the chemicals used to process circuit boards," Lofty adds. "We typically work with ppb [parts per billion] criteria. If you walked all around the world and measured it in inches, that's ppm [parts per million]. So you can see that we're working with extremely demanding orders of purity and that quality per yield is critical. If you can get 90 boards instead of 75 out of the same run, and you can sell them at $5,000 apiece, that's $450,000 instead of $375,000. That's profit. Quality and consistency really does equal dollars."

Competitor Dispute Drives Lab Automation

Working at parts per billion levels usually also requires state-of-the-art equipment at all stages of the process. However, with the exception of a few instrument controllers, Electronic Materials's labs were not automated. Their rigorous attention to detail compensated for the time-consuming requirements of managing the system manually. However, the need for a Laboratory Information Management System (LIMS) that could provide fast access to quality data was growing to a head.

While Lofty believes that scientists are very good about writing down ideas and tests, he also states that they can't do it on paper any more and maintain high levels of productivity. Shortly after joining Electronic Materials in 1989, he zeroed in on the lack of automation in the lab. During a strategic meeting, he argued vociferously in support of a LIMS, saying that they had to have a LIMS in order to have a competitive edge. He ended up being put in charge of the project.

A dispute with a competitor over intellectual property created the final argument in favor of a LIMS. "We needed to prove that we created the product first," reminisces Lofty. "But nothing was automated at that time. So, in a 15-foot room, knee-deep in lab notebooks, three secretaries spent two months going through lab information, copying it, and sending it to Legal before a single five-year-old lab notebook was found that contained the original entries. We won a multimillion dollar lawsuit, but it was clearly time to get a LIMS."

Electronic Lab Data Expedites ISO 9000 Implementation

After a substantial review and analysis, the LIMS was subsequently installed in 1992. Before the end of the installation, the division was chartered to implement ISO 9000. "Since the early 1980s, we've taken numerous TQM and SPC classes," Lofty observes. "But unless you have a means of doing it easily, nobody's going to draw hundreds of charts. Now, with information in an electronic format, we could examine what happened, why it happened, and determine how to fix it."

The LIMS helped Electronic Materials expedite the ISO 9000 implementation because the electronic format of the laboratory information enabled them to provide Statistical Process Control (SPC) on all their materials, not just a small percentage of the product.

"With ISO 9000," he continues, "where before we thought we had a tight system in place, by using lab-built templates, we learned a lot about our processes. We had a big book of procedures and a big book of specifications. The very first test for a product that we did—a 100-ppb chloride in ICP—test number 47 as a matter of fact—when we looked for the procedure, we couldn't find it in the LIMS. We found out that we do the test by titration. When we asked about it, the response was 'everybody knows we do it by titration.' We've now had the technical document department write test number 47 correctly."

Using an electronic format for laboratory information means that Electronic Materials is now able quickly and easily to confirm procedures, specifications, and methodology for more than 480 procedures. Documents can now be called up at any workstation in two keystrokes. In addition, 95 percent of their sample data can be located in less than five minutes.

"Before, documents were in a manager's office—a holy shrine so to speak—and nobody read them," Lofty recalls. "We turned the office upside-down. Documents were created in WordPerfect and fed into the information system. Reading the procedure is now part of the job responsibility that resulted from an ISO 9000 analysis."

Refining their processes means that Electronic Materials no longer uses paper specifications or paper procedures. This is particularly advantageous in a clean room environment where an operator has to put the paper in a plastic sleeve and then put on a bunny suit before entering the clean room—a time-consuming process. "If they grabbed the wrong procedure by mistake, out they go to start all over again," he adds. "With electronic specifications and electronic procedures, they just pull the information up on the screen when they get inside. In addition, electronic information has shaved time off getting procedures out to people and there doesn't need to be reminders of new procedures if they only access the latest version."

Real-Time Availability of Procedures and Results

When asked about the benefits of automating their quality and lab data, Lofty immediately highlighted improvements resulting from real-time availability of procedures. "We've eliminated having material being tested to the wrong testing procedure, and we went to zero in material being tested against the wrong technical specifications. We're now working on taking rework down to zero."

In addition, every parameter that's tested is put into the SPC format and tracked. "Every analysis is SPC'ed, the water, the components, whether it's in control or out of control, in spec or out of spec. Running a spec is not sufficient. We have to be consistent. So we've adopted global SPC control. When results are put into the instrument, it's a two-button operation to get an SPC chart that is built over past results. This gives us real-time analysis that can pinpoint, for instance, when water specs are out of control, enabling us to determine if, say, a valve cracked open. If the same incident happens three times, then corrective action kicks in for valve replacement."

Electronic Materials achieved ISO 9001 certification in August 1993. The benefits continue to accrue. "It used to be that R&D was always getting beat up by Marketing because it used to take 15 to 18 months to get a product out. Now it takes 5 months and Marketing says they can't do the market research in less than 6 months. With our new laboratory and quality systems, we're making their lives more interesting. In fact, we've reduced the product introduction cycle by cutting the original R&D cycle time by more than one half to one third."

A Change in Corporate Culture

As processes changed and new capabilities were added, there was a corresponding change in the corporate culture. "We were quite surprised by this because the labs have always operated reasonably well," Lofty comments. "But their processes were time consuming. Guys from production used to bring in samples and wait until the lab could run an analysis."

"Now," he adds, "those same production staff have received end-user training and can use the LIMS to log samples, receive test results, and get the chemist's approval of the results. This has eliminated forms, too—we no longer need Sample Analysis Request forms. In fact, it took longer to fill out the form than it does to log in the samples and have LIMS check all the tests. The process now takes less than five minutes and all the chemist has to do is review the results against specifications."

As originally conceived, Electronic Materials planned to train two people to design experiments and SPC charts. However, after the LIMS was implemented and the people were trained, one of them ran 100 trials on an experiment and discovered he could have had more information on 16 trials if the experiment had been designed differently. This discovery had a significant impact.

Initially, it wasn't easy to get people to move past being ruled by the procedure to creating and ruling the procedures. But Lofty drove decision making down to the bench level. When that happened, staff started red-lining procedures so fast that management couldn't review the changes in a timely fashion. So bench level staff were empowered to approve their changes. This shift from reactive to proactive procedure control is now self-perpetuating.

According to Lofty, "The key is to get the manager out of the process and away from signing off on specifications. It used to be that six people had to sign off a specification. Now only one person signs and approves it and everyone else has review responsibilities. We've given empowerment to whoever does the work, and sign-offs now take place in the system."

Self-Perpetuating Change

With procedures and information available electronically, it was just a matter of time before the data were used on a global scale. "We recently connected all our systems at all our sites using a worldwide satellite network," Lofty enthuses. "This means that our U.S. chemists can immediately see results from film manufactured at the Kodama, Japan, site. This means that customer input in Germany can get reviewed in California where the product is designed and produced."

Lofty is adamant that it is a marketing advantage to be able to provide nonsubjective evidence of product quality through SPC. But the changes that the labs at Electronic Materials implemented provide much more than that. Close scrutiny of the flurry of activity that initiated and accompanied the implementation of ISO 9000 reveals that Electronic Materials has not followed the standard to the letter. Instead, they took the intent of the standard and launched a stellar continuous improvement process built upon self-perpetuating change.

If the key to building a business is continuing to improve the product, and if the key to improving the product is improving the processes that surround the product's creation, then Electronic Materials is helping to ensure that Morton International is going to be around for another hundred years. Proving product ownership is just one part, albeit a critical part, of the cycle. Using state-of-the-art technology to streamline the processes surrounding product development, improvement, and production can also provide a strategic competitive edge.

Lofty's evolving role first as a leader, then as a facilitator, underscores the type of leadership necessary to implement the changes that organizations will need to undergo in order to remain competitive in the global marketplace. But he refuses to take much credit for his role. "Our corporate culture was moving that way—my management style fit into it perfectly." Despite his modesty, his energy and vision were key to not only instigating the change but ensuring its success.

Electronic Materials's ISO 9000 implementation experience proves that the old adage "if it ain't broke, don't fix it" no longer holds true. Just because a process works today doesn't mean it will be adequate to address future challenges. Reengineering key areas of the business, such as the lab, using quality management tools, such as ISO 9000, will become a continual process. Wise organizations will follow Morton International's lead and anticipate change, and in doing so, precipitate change.

A

Interpreting ISO 9000, Element by Element

Explanation of Elements and Requirements

Following is an explanation of ISO 9001 in which the requirements within each section are reviewed in depth. You will no doubt find that in some instances you have already documented certain procedures or that you are already performing inspections which meet the ISO 9001 criteria. Congratulations! There will probably be several areas, however, such as internal quality audits, that you will need to implement. Remember that implementation is the key: you won't pass the third-party audit unless you have not only written down what you do, but you must have implemented what you've written down. This is proven with a document trail.

This appendix is organized by first laying out an interpretation of what the clause means, followed by a commentary on how the 1994 revisions changed or affected or strengthened the particular clause. It's best to have a copy of the relevant ISO 9000 standard with you when reviewing the explanation and suggestions below.

4 Quality System Requirements

4.1 Management responsibility

4.1.1 Quality policy

What's required is a written quality policy signed by the head of the site. One page is fine. The more succinct the better. The policy, however, is part of a larger document, typically called the Quality Policy

Manual. It's usually about 25 to 35 pages long. In it, you'll address each of the 20 elements, usually with one to two pages apiece.

The requirement to ensure that the policy is understood throughout the company means that the auditors will expect to see the policy posted on the walls and that employees can recite it without error. Your employees will also need to be able to explain how it applies to them. For this reason, it's a good idea to conduct internal ISO training and make it part of your new employee training. (More on that later under 4.18 Training.) Also, even if you're one site belonging to a larger company, the policy should be signed by the person in charge at that site. The scope of your ISO 9000 certificate will not extend beyond the site, or sites, you define. If the president of the company doesn't oversee operations at your site, and won't be running your Management Review Quality Meetings, then the certificate should be signed by the person who does.

Note the word "objectives" in the first sentence of this element. This means you must have goals for process improvement. If you build in continuous improvement as one of your goals, you're meeting these criteria.

4.1.2   Organization

4.1.2.1   Responsibility and authority

Defining responsibility and authority requires more than an organization chart, although the organization chart is often included in the top-level Quality Policy Manual. Be warned, however, that including the organization chart is including a document that changes frequently—particularly if your organization is dynamic. Within your Quality Policy Manual, you need to explain who is responsible for the different elements. For example, depending on how your company is organized, Shipping may be managed by a Shipping Supervisor, a Distribution Manager, a Warehouse Manager, etc. There are many different titles for this function. The person who approves the product for shipment is the one who needs to be listed. Of course, so does everyone above that level since they approved the approver for the task.

Despite this section being called "Management Responsibility," it goes beyond management to include anyone responsible for managing the work that affects quality regardless of the employee's level in the company. If something goes wrong, who reports the problem? Who resolves it? That's what Sec. a is asking. Is it the responsibility of the line worker in Assembly to report and record product quality problems? Then you need to state this in the policy manual. Same thing for the other areas. Who is the responsible authority? If you make it the manager every time, he or she is going to be a very busy person.

That's why it makes sense to define the responsibilities that are delegated down and list their titles and specific quality responsibilities in the Policy Manual. This is not quite a job description, but rather a list of the tasks they do that affect quality. Other tasks are not under consideration, but if you include them you will be audited to those tasks too. To keep it simple, stick to the quality tasks.

Including titles and responsibilities in the Policy Manual can mean that there is a four- to five-page section just on quality management responsibilities alone. Some companies have put this in the second level of documentation, the Procedures Manual, but more and more auditors expect to see it in the top-level Policy Manual.

The 1994 revision added the requirement that this information be documented. It also broadened the definition of a product nonconformity to include any nonconformities relating to product, etc. Remember that with ISO a process nonconformity is as big an error as a product one.

4.1.2.2 Resources

This section can be bundled in with 4.1.2.1. Those personnel who are performing verification activities need to be defined in the responsibilities section. In addition, it should be the manager of the person performing verification who is responsible for ensuring that they have the tools and training to do the job. If the manager is not also responsible for identifying what needs to be verified, then the person who is responsible must be identified. The 1994 revision eliminated an entire paragraph from this area which defined verification activities as including inspection, test, and monitoring, etc. By eliminating this clause, the definition of what a verification activity is has been broadened. Thus, any time a check or verification is performed, the auditor has the power to state that the activity is a verification and falls under this part of the standard. This was given more strength because the scope of personnel was expanded to include personnel performing management and performance of work as well as the verification activities associated with such work. The final phrase regarding internal quality audits was also added, no doubt to ensure that these were done before the initial assessment audit as well as that such internal audits continue to be done on a regular basis thereafter.

4.1.2.3 Management representative

The management representative is the person who spearheads implementing the standard and ensures that it is maintained after registration is achieved, as well as being the contact with the third-party registrar and assessment body. Typically known as the ISO Project Leader, the person who performs this function can be anyone from a

Quality Manager to a Controller to a Vice President. It matters little who it is; you can even hire a new person to fill the function.

Generally, the management representative responsibilities are layered on top of one person's existing duties. So if anyone has the right to complain about how much time ISO is taking away from their other responsibilities, this is the person! But the management representative or ISO Project Leader is usually the one most enthusiastic about the standard (it's almost a requirement for success) and often plays the role of internal ISO evangelist within the company. Of course, the management representative must also be listed in the Policy Manual and his or her ISO responsibilities as the management representative must be defined separate from other quality responsibilities under any other function he or she performs.

The 1994 revision updated the phrasing in Sec. a so that not only must the requirements of the standard be implemented and maintained but established in accordance with the standard. A fine point, but one that the Technical Committee felt needed to be made. In addition, Sec. b is completely new to the 1994 version. So, extra merit has been given to reporting the performance of the system, which means that this is expected to be discussed at the management review meetings that must be conducted at least annually to show that your company is giving appropriate attention to the quality system.

4.1.3 Management review

This is the first of the required procedures that you'll need to implement if you're not already having regular management reviews of your quality system. The purpose of the management review is to ensure that the quality system is functioning and being followed. The review that is being highlighted here is not one that is conducted by one person. Rather, this requirement is satisfied by an annual meeting chaired by the top manager at the site (often the one who signed the policy statement) and during which quality issues are discussed. The procedure should state that the agenda will include discussion and assessment of internal quality audit results (4.17), incidences of product and process nonconformances and corrective actions, and a review of areas marked for process improvement. Quality objectives should also be covered and, if you assign action items during the meeting, a review of those action items should be on the next agenda. Root causes and trend analysis of nonconformances should be discussed as well as issues such as reject rates, rework rates, defect types, and customer complaints.

During ISO 9000 implementation, many companies find that an annual meeting is not enough, and that there are other quality meetings that probably should be held. So, management reviews, like management responsibilities, are not limited to upper management. Yes, upper management must have an annual meeting, but you need to address quality issues more frequently than that. Typically you'll have an ISO Steering Committee that meets weekly, biweekly, or monthly. Formalize the agenda and write a short procedure for that meeting. What about product quality discussions in Manufacturing, Assembly, or Distribution? Are those or any other areas of your company conducting regular reviews to discuss rework, scrap, or nonconforming product issues? Again, formalize those meetings and write a procedure for each. You'll probably be pleasantly surprised at the number of product quality meetings that are already occurring but aren't documented at your company.

Now, take a look at the last sentence in this element regarding records. It's an important one. Wherever it crops up in the standard, you are required to keep appropriate records. In the case of management reviews, this means meeting minutes and copies of the agenda. It's probably a good idea to start a new quality meeting file that contains minutes and the agenda. The auditor will ask to see evidence that the meetings occurred. If you start a new file, you'll have one readily at hand. Don't scramble for it during the audit.

The 1994 revision rewrote this clause almost in its entirety. Why was the Technical Committee so dissatisfied? Apparently, management reviews weren't taking place at a high enough level. The previous wording also said that the reviews could take place at appropriate intervals; now the text reads at defined intervals. So, you must be specific about how often these meetings take place and be prepared to show the auditor the document which defines this.

Overall Summary of the 1994 Revisions to 4.1 Management Responsibility

The role of the management representative is given teeth in this section, by requiring that this person have executive responsibility and that he or she reports on the performance of the quality system. This means no more off loading of ISO quality responsibilities onto a supervisor, but that the top-level management has to be directly involved. It also requires:

- That the quality policy be more than a platitude, and be closely aligned with the organization's mission statement.

- Nonconformities and product quality problems to be reported and acted upon have been expanded to mean problems with quality processes and the quality system as well.

- Personnel responsibility must now be documented as well as defined; this means that job responsibilities must be spelled out in writing.

- The subhead entitled "verification resources and personnel" has been modified to read simply "resources." The definition for verification activities is eliminated; now it simply states that resources should be available to do the job, not just verify the job. By changing the title and expanding the meaning of verification to include providing the tools and staff to do the job right, there could be far-reaching implications depending on the auditor.
- Continuous improvement is introduced into the standard with the statement that "reports on quality system performance are to be used as a basis for system improvement"; this means that continuous improvement must be a topic of management reviews.
- Note 5 introduces the first mention of external auditors into the standard and indicates that the management representative should be the one to liaise with them.
- Management reviews will now take place regularly as opposed to when appropriate; this will require specificity of some kind. The Note on what happens during a management review has been eliminated; this means that the scope of the review is up to the organization, but will probably require more depth.
- Significantly, the 1994 revision now states that the management representative shall review the quality system, not that the quality system shall be reviewed by management. So while the annual management review may still be sufficient, the representative should review the system more frequently and have the power to call additional meetings as necessary.

In summary, this section points to greater involvement by upper management, who will need to ensure that they are providing the resources to do the job, not just lip service, and that continuous improvement must now be an issue, not just nonconformance prevention.

4.2   Quality system

4.2.1   General

4.2.2   Quality system procedures

4.2.3   Quality planning

Documenting your quality system means having a quality system, a quality plan, and a quality manual. Start by preparing a Quality Policy Manual and documenting your procedures, work instructions, specifications, forms, etc. Typically you will have three levels of documents: a Policy Manual, a Procedures Manual, and a Work Instructions Manual, a pyramid of manuals which define the organization's quality process in more and more detail. Forms and specifications are also controlled but usually kept in separate binders. The effective implementation of the quality system, including both documents and actions, is monitored through internal quality audits (4.17) and verified through external third-party audits.

While addressing the Notes is not required per se by the standard, they are critical to your success in getting registered. The requirements in Sec. a are usually rolled into document control (4.5) such that a section regarding writing and maintaining the quality documentation needs to be developed. Section b is also spread through several procedures since it affects those in which measurement and test equipment are used; control points where verification inspections take place; production plans regarding staff and materials on hand; and required skills or training to perform the functions. Updating quality techniques as laid out in Sec. c can be rolled into one of your quality meeting procedures or it can be driven at the department level. Even if you haven't determined how to measure something as outlined in Sec. d, you need to identify the need for that measurement and state this at any control point in your process where the measurement would take place. Standards of acceptability must be spelled out as stated in Sec. e, which will occur at any control point, verification, or inspection in your procedures. Compatibility of product issues with procedures and documentation should be documented per Sec. f. In other words, are you doing what you say you're doing, again, underscoring the requirement for accurate documentation. Section g shows how the standard's requirements interlock with each other—as indeed your own documentation will—by cross-referencing 4.16 Quality Records.

What you'll need to do after you've put together your quality documents is to go back and scrutinize them against the quality system requirements listed above. You'll need to ask whether all the quality system issues are adequately addressed in your documentation and if you have accurately identified and written down where each of these issues occurs.

Overall Summary of the 1994 Revisions to 4.2 Quality System

There are new subsections and more detail has been added to this section. Previously, there was simply a paragraph concerning the quality system with seven Notes; now there are three subsections, including procedures and planning, and the Notes have been incorporated into the requirements. Most significantly, a quality manual is required, and that manual shall refer to the procedures and define the structure of the documents used in the quality system, meaning that this should no longer just appear in the work instructions. The following revisions should also be acknowledged:

- Note 6 points to ISO 10013. This means it would be a good idea (since the auditors are going to refer to 10013) to revisit the quality manual to ensure that it matches up with the guidance given in ISO 10013.

- The number, length, and format of the procedures, methods, and personnel skills and training are now acknowledged to need to be appropriate to the organization. This gives the organization the right to create concise documents and use unskilled staff (though giving them on-the-job training) when the job is straightforward.

- Note 7 suggests that procedures reference work instructions; this means that there should be an additional section added to the beginning of each quality procedure document that refers to related quality work instructions documents.

- The incorporation of the Notes makes quality plans a requirement, not just a suggestion. Essentially, if the Notes were addressed before, the organization is already performing the requirement; however, the revision gives new teeth to identifying suitable verification at appropriate stages. This means that verification should probably take place more often than at the beginning and at the end.

- Note 8 gives an option in preparing quality plans in that they can be presented in a matrix format.

In summary, there is not much change to this element if the organization had previously addressed all the Notes. The main changes center on ensuring that the quality manual addresses the standard and that planning which incorporates quality issues occurs. A documented quality plan is now, however, required.

4.3 Contract review

4.3.1 General

4.3.2 Review

4.3.3 Amendment to a contract

4.3.4 Records

The words "contract review" are a British term for review of customer orders, not a review of legal contracts. In essence this requirement covers order processing, but also applies to outgoing proposals as well as incoming orders. This is because Request for Proposals

(RFPs) generate work orders. Since the end result of a product's quality begins with an accurately defined order, you'll need to write down your procedure for accepting orders and how those orders are reviewed.

This section doesn't particularly address the sales department. However, if contractual requirements are an issue, then the order may require review by a Sales Manager, Financial Controller, or other higher-level manager before it can be released. Even so, this requirement is directed primarily at what happens after the order is signed. If the order processing department is not located at your site, and hence outside the scope of your registration, you still process a request that fulfills an order. Therefore, the order processing department that sent you the order is your customer (or purchaser) in this instance, and the person in Manufacturing or Distribution or wherever who first initiates order fulfillment is the one performing order processing at your company.

You will need to determine if your products are standard or nonstandard. Standard products are, of course, the same every time they are built. Nonstandard products are those built to custom specifications. If you process nonstandard orders, or if all your orders are custom-built, then you need to define the approval process for ensuring that each set of specifications (i.e., each order) can be met. The most important thing to remember is that you need to document the approvals, dates, and any revision numbers. In the case of a verbal purchase order, fax a copy of the order to the customer and request that the signed order be faxed back to ensure that your organization receives written confirmation.

Again review the requirement for records. What the auditor is looking for is evidence of the customer order and how it was processed. If, for instance, you use a Work Order with accompanying Travelers, any changes to the Work Order must be transmitted to the order processing function. Therefore, there needs to be a contract or customer file that contains all changes and approvals to the Work Order.

The Note that is included in this section which refers to coordination with the purchaser's organization does not necessarily mean that you need to involve more than the Purchasing department at the customer site. Typically, you are already requiring confirmation of an order before you fulfill it, and when you do so, you probably review the order requirements. ISO 9000 just wants you to write it down so that you always review the same issues every time. It does not, however, lock you out of putting through rush jobs or changing the order after it has been initiated. You merely need to insert a paragraph or section in your procedure or work instruction about how you handle rush orders as well as ensure that whenever changes to an order occur they are documented.

Overall Summary of the 1994 Revisions to 4.3 Contract Review

This section is now broken out into four subsections, giving greater emphasis to each issue. The definition of contract has been expanded to mean any customer order, and amendments to orders must now be identified. In addition, the following changes should be noted:

- Orders should be reviewed for completeness before or when they are accepted; there was no specific time constraint previously.

- If the order is not in writing, requirements must be agreed upon before acceptance; this means that some indicator must be determined to ensure that this activity takes place, such as faxing a verbal confirmation of the order; the auditor will look for documented proof even though this is not directly stated.

- While the intent is more or less the same, because there was significant rewording to the need to resolve differences in order requirements and what the supplier can deliver as well as ensuring that the supplier can fill the order, it is a good idea to revisit who reviews and signs off orders before they are input and how special and custom orders and other requirements are addressed before order processing.

- Handling of amendments is now specifically addressed and organizations are now responsible for identifying how these are made and transferred to appropriate functions. This will require documentation, both in the form of quality procedures and work instructions and in backup records that support the performance of amendment circulation.

- While the previous Note suggested that activities within the organization be coordinated with the purchaser, the revised Note 9 suggests that communication or interface channels be established with the customer. This subtle change implies that staff should know who to call at the customer site when there is an issue with an order.

In summary, more emphasis has been placed on ensuring that the product is delivered as ordered. To ensure this, the front end of the ordering process and the handling or communication of amendments to an order have come under greater scrutiny and will probably require more detail in the documentation.

4.4 Design control

4.4.1 General

Section 4.4.1 is fairly self-explanatory. Basically, you need a design procedure. You do not, however, need to define the creative process, only the handoffs between the various design stages. The rest of the design control section goes into detail on what needs to be covered in this procedure. One thing to remember is that the auditors have a very irritating habit of asking "But how do you know?" They will want to know how you can confirm that design changes happened, where those changes were documented, and whether the end result matches the original intent.

4.4.2 Design and development planning

Once again the word "plan" crops up. You're probably already planning new products with weekly design committees who discuss the project status and assign action items. You need to formalize those meetings, use an agenda, keep minutes, and document the action items. ISO 9000 is concerned with the methodology you employ throughout the design process to develop and verify the product or service being designed. The standard wants you to consider and document what activities should and do take place, and that personnel assigned to those tasks have the means to perform them satisfactorily—something you'll need to prove too, usually by training records. The standard also wants you to define the interfaces necessary between all the groups involved with design throughout the process. This means defining when marketing should be involved, when quality should be involved, when management should be involved, etc. If sign-offs are necessary, there needs to be written evidence that the sign-offs occurred.

During this phase you will want to establish design verification points that will be used later in the design process to confirm that the product conforms to the original plan.

4.4.3 Organizational and technical interfaces

This is a new section, but not a new requirement to ISO 9001. It was previously a subset of design and development planning. Because it was separated, it has been given greater presence in the design process, meaning that auditors will look to ensure that the input of all functions involved in the design process, from marketing to quality assurance to manufacturing, is documented appropriately.

4.4.4 Design input

Another aspect of documenting interfaces is determining your design input requirements and having them reviewed for adequacy. We've all heard of design departments that have developed products without the input of marketing, sales, finance, quality, etc. A product gets built that costs more than the market will pay or has nifty features that the market doesn't want. These products are not adequately covered by design input since they were created in a vacuum. ISO 9000 seeks to assist you in optimizing design issues by planning your design and/or development so that the end result matches the specifications. By having more than one group review the design, it is more likely that problems with the design will surface. For example, it may be that marketing has asked for the product to be designed at a cost that can't be met with the specified materials. That's a conflicting requirement that needs to be discussed and resolved.

During this phase, the product performance requirements and specifications are typically defined followed by the identification and resolution of all design ambiguities.

4.4.5 Design output

The goal here is to ensure that you built what you said you would. If there are acceptance requirements that must occur before the product can be released, such as inspections and tests, those need to be specified as well. If regulatory requirements or safety issues are a factor, they must be documented and defined prior to release. This requirement can be quite a hurdle for companies developing complex products such as software. Every change to code must be documented, otherwise how do you know the change was made or why? Even if you're developing a straightforward product such as a common interchangeable plastic part, there may be strict requirements for color, strength, tolerances, and heat resistance. Those criteria need to have been addressed during planning, listed in the design input, and met by the design output. Design output is where you write it all down and can prove that you met the requirements. Note that the requirement to meet the design input requirements is repeated twice.

Design output entails committing the design to hard copy of some form, such as drawings, calculations, lists specifying the designated product materials, etc. Often a prototype is built during this stage.

4.4.6 Design review

This is a new section for the 1994 version of ISO 9001. It was covered by default in the previous version, although apparently not enough organizations were sufficiently documenting this rather crucial step, thus warranting the addition of a specific requirement to do so. Most organizations ensure that they meet this requirement by developing flowcharts of the design process and reconvening portions of the design team at various intervals to review the progress as well as the results. Minutes of those meetings, and the resolution of action items that result, will satisfy this clause.

4.4.7 Design verification

Throughout the design process the design should be verified at different control points. It is usually the responsibility of the design planning function to define where those verification control points should occur and what they will be. It requires that the people who perform verification know what they're doing and is a double-check that the design output meets the design input. You'll note that this particular statement occurs twice previously. That should be a red flag to you that this requirement is critical. Hence, it will be closely examined during an audit. Regardless, it should be closely examined anyway and the design validated. Verification might include design reviews (part of 4.4.6) as well as alpha and beta testing. Interestingly, much of the Note here was previously part of the general text. This "downgrade" means that such steps are not required, but suggested, and should be performed if appropriate. Every industry has different types of verification activities, and ISO 9001 does not seek to dictate what those activities should be. However, this clause is a new one, so while the Note may take the pressure off certain requirements that need to be fulfilled, the fact that design verification must be performed is given more emphasis.

Design verification includes activities such as qualification tests. If the product is software, such as a Laboratory Information Management System (LIMS), the software must be verified to ensure that bugs are eliminated. Beta-testing the product internally or alpha-testing the product at a customer site often occurs during this phase.

4.4.8 Design validation

This section is also new with the 1994 revision. As you can see, it provides a slightly different emphasis from what was previously included. Its intent is to ensure that the end product performs as expected when being used. The various Notes point out the criteria that should be addressed during this stage. This emphasis makes sense in light of the greater regulatory and validation issues faced by companies in a variety of industries. ISO 9001 simply seeks to align itself with other industry and government requirements.

If beta- or alpha-testing has not yet occurred, and is appropriate to the product, it is performed during this stage.

4.4.9 Design changes

Throughout the design process there should be room to modify or change the design. Designing a product should be a flexible process that can take advantage of, or react to, changes in the market. Changes require documentation and approval of the change. This is not meant to hinder the design process, but rather to ensure that the design is not changed without authorization or accurate documentation that defines the nature of the change. This requirement can be covered by a short paragraph in your manuals, but adhering to the change requirement can become a documentation-intensive task. Because of this the following may occur: There may not be as many changes to the product, changes will be more carefully considered, and more emphasis will be placed on the design planning phase. In addition, critical aspects of the design will no longer be in the head of a key employee who could walk out at any time leaving you with nothing. Now, your product design belongs to the company, not an individual.

Design changes may be necessary because the design team runs into manufacturing difficulties, or better technologies are found, or the customer requirements change. When design changes are necessary, the process begins again at the point where the change is applicable.

Overall Summary of the 1994 Revisions to 4.4 Design Control

This section was streamlined throughout. In addition, it is now included in 9002 and 9003. It was added to align the numbering of the two standards. As such, it is recommended that the numbering of existing 9002/9003 documents also be renumbered to correspond with this revision.

The requirement for documenting the design steps was added. In addition, the responsibilities in the design and development planning phase now need to be defined, as well as ensuring that there are the resources at hand to accomplish the tasks set forth. Additional emphasis was placed on the organizational and technical interfaces, so auditors will look to ensure that you meet these criteria. This means that just because a function or department does not fall under the specific title of "design," such as sales and marketing, because those departments provide input into the design process, their input must be documented appropriately.

Sections for design review and design validation were added to ensure that the end product performs as expected. No "oops!" allowed. Interestingly, a clause requiring the review of applicable statutory and regulatory requirements during the design input phase has now been included in the standard. No specifications are included, of course, since this standard is a generic boilerplate for all industries, but now industries selling into the environmental, pharmaceutical, medical, etc., markets that must meet government regulations must ensure that those regulations are addressed during the design phase and, of course, documented.

4.5   Document and data control

4.5.1   General

If you do not have a document control procedure already, you will need to implement one. However, your present procedure may not address all the requirements of ISO 9000. Use your document control procedure to define the procedure structure, how documents are maintained, how often they are reviewed, what arrangements you have made to ensure that only current documents are used, and what documents are considered controlled. Forms also come under document control, as well as specifications and measurement charts, as appropriate.

Documents can also be customer specifications, drawings, and the Approved Vendor List—anything related to the quality of the product or system processes. The key issue regarding documents and data is to ensure that they are approved and dated, and that there is revision control.

4.5.2   Document and data approval and issue

Before a new version of a document can be issued, it must be reviewed as stated in the requirement, and you need to have proof of the approval; that is, a document trail that can be audited. You need to ensure that only the current revision is being used. This issue can be addressed by placing a red stamp on the master document so that all others are obviously copies.

The subsections particularly refer to your quality procedures and work instructions. These must be not only current, but accessible to staff performing the functions in the documents so that if they have a question, they know where to find the answer. One of the things that auditors look for is whether posted documents are controlled (they should be), that only the current version is being used (it must match the master controlled document), and that staff performing the function know where the documents are (a simple issue that has tripped up many a firm that hasn't ensured that *every* employee knows where to look).

Most companies design a Document Control Matrix which lists the types of documents that need to be controlled and how long they need to be kept. It's a good idea to check with your legal department to determine which documents, such as employee records, need to be maintained for a specified period of time by law. You'll need to know what those timeframes are, write them down, and adhere to them. In addition to the actual performance of controlling specific documents, within the applicable procedure you need to ensure that documents—such as Work Orders—are reviewed and signed off by approved personnel. This is a control on a different level, but one that is extremely important to document. You will be audited against the required approvals and the auditor will expect to see evidence throughout your procedures, particularly at control points, inspections, and verifications, that the proper authority signed off the document.

4.5.3   Document and data changes

The first paragraph addresses changes to your quality documents; that is, you cannot change a procedure unless the original authority reviews and approves the change. In addition, they can't perform such reviews and approvals in a vacuum, but must have access to the appropriate information for making a change. This means that the person who wrote the procedure should be the one who revises it, and if it's not the same person, that person still must be notified when someone else wants to make a change. Many companies create an Approvals Matrix that lists who has the authority to make changes to a document. Be aware that although a Vice President can change a document without running it past the Supervisor who wrote the procedure, if that Supervisor is the approval authority for the procedure, he or she must be notified of the change. Be careful! While it might seem like a simple idea to just have the Vice President be the sign-off authority for all the procedures, you'd swamp him or her in paperwork if you do so. Thus it is a good idea to delegate down and make the departments responsible for maintaining their procedures, not upper management.

If you haven't created a master list of the controlled documents, as mentioned above, you need to be able to identify current revisions in some way. A master list is the simplest and most straightforward answer.

Finally, think of numbering your documents in the same way that a software manufacturer numbers software. You get version 1.0, then 1.1, then 1.2, etc., until a major revision is done and now you get version 2.0, etc. In your procedures, it's easiest to think of using 1.1, 1.2, etc., as indicators of minor changes and 2.0, 3.0, etc., as indicators of major changes. If you determine to perform quarterly internal audits, then it might be simplest to do a complete revision after correcting the results of each audit and using the fractions as a means for the individual departments to alter their procedures between audits. Do whatever makes sense. But what you do, and how frequently you revise your documents, needs to be written down and become part of your procedure. The auditors aren't looking for endless detail; you can address this requirement with a simple paragraph.

The final sentence in this clause is interesting. Most companies include a separate page with each procedure that lists the changes, including the date and what was changed. This could get extremely tedious in a few years when the changes may be longer than the procedure. One company, DisCopyLabs of Fremont, California, has ensured that they address this issue in the actual document itself by striking through all previous text that has been eliminated and placing new or revised text in italics. This enables the person reviewing the document to quickly see at a glance what was added and what was eliminated.

Don't forget to identify obsolete documents that are retained for legal or archive uses. It's a good idea to write OBSOLETE on the document to ensure against use.

Overall Summary of the 1994 Revisions to 4.5 Document and Data Control

The first sentence in this revision jumps out with a new requirement to have a system for controlling external documentation that affects quality. This means that reference manuals and materials must also be defined, listed, and controlled as part of the internal quality document handling system. In addition, the scope of this section has been expanded to emphasize the control of data as well as documentation; however, it accommodates the real world by stating that this information can be in any form, from hard copy to electronic media to whatever comes up in the future. The following issues are also raised:

- While it is no longer implied that obsolete documents must be destroyed, if they are retained, they must be identified as such; this means that obsolete documents should probably be marked in some way.

- The requirement to reissue documents after a practical number of changes has been eliminated; this leaves it up to the organization to determine when a reissue should occur.

In summary, little has changed with this section other than to address the issue of controlling external documents that affect quality. This means that organizations will need to set up a system which identifies and controls these materials. A look at the existing document control procedure may warrant a corresponding change to address that issue as well as document obsolescence and reissues.

4.6 Purchasing

4.6.1 General

Sometimes the simplest statement is the most deceptive. Yes, you need purchasing procedures, but as you'll find in the following sections, those purchasing procedures should include three distinct elements. As you are no doubt beginning to discover, your existing purchasing procedures probably don't address all the requirements.

4.6.2 Evaluation of subcontractors

You can't just put a vendor on dock-to-stock status, you need to have documented workmanship criteria for achieving that status—and not documents from the vendor either, but documents you create based on inspection criteria you develop. That done, you need to review vendor products periodically to ensure that they are still up to snuff. In addition, sometimes the Shipping department maintains a list of approved vendors such as freight forwarders. This is a purchasing function, even if it is not performed by the Purchasing department. Therefore Shipping needs to have a purchasing procedure of some kind as well, or deliver the maintenance of their Approved Vendor List to Purchasing. Yes, an Approved Vendor List is something that the auditors will look for, but it doesn't necessarily have to be a controlled document, just an up to date one.

Most companies address this requirement by formalizing or creating a procedure for vendor assessment and monitoring. This procedure needs to go beyond buying by price to weighing the quality of what you're buying. You can "grandfather" in vendors that you have been buying from in the past, but you will also need to implement a formal vendor review or monitoring process. It may mean that when Manufacturing discovers problems with raw materials, Purchasing is copied on the Return To Vendor Report. One way or another, Purchasing needs to be able to make informed decisions on vendor status.

An initial assessment, perhaps a probationary period of three to six months before a supplier can go on "approved" status, needs to take place. You need to establish criteria against which all vendors can be assessed, as well as a means of monitoring their performance. Write down how a vendor gets on the Approved Vendor List as well as how the vendor gets taken off the list. Steps that you might put in this procedure include the information below in Purchasing Data.

4.6.3 Purchasing data

Again, spell it out and keep track of what you've done. You need to be very specific on your purchase orders, particularly if you are ordering a product or raw materials to an exact specification. And, if you use the same part number but switch suppliers, remember that you need to assess the new supplier for that part number even if you already order other product from that supplier and the supplier is on the Approved Vendor List. Product quality can vary and you need to assess each product separately. If appropriate, include drawings or specification numbers on the purchase order.

4.6.4 Verification of purchased product

4.6.4.1 Supplier verification at subcontractor's premises

4.6.4.2 Customer verification of subcontracted product

Once you've cut the purchase order, what controls do you have in place to ensure that the product you ordered—especially product that is going to become a component in your own end product (i.e., raw materials)—is the one you want? Incoming inspections are the typical answer. However, you need to question whether incoming inspections performed by Receiving are adequate. After all, Receiving inspects for damage and looks to ensure that the part number on the purchase order matches the part number of the product being received. Receiving does not typically inspect the quality of the incoming material. Would it be more appropriate for someone from Purchasing to perform the incoming inspection? Or someone from Quality Assurance? Some ingenious companies such as Glaxo Wellcome place analytical instruments, such as Near Infra-Red (NIR) detectors on the receiving dock to test incoming materials upon receipt. If the material isn't of acceptable quality, it never gets inside the building, but is returned to the vendor. Just because Receiving hasn't typically performed such inspections doesn't mean they can't be trained to do so. The important issue here is not who does the inspection, but that the inspection be done. The second part of this clause even accounts for customer verification of materials that are to be used in their products. This, of course, depends upon the particular customer-supplier relationships you have. Alternatively, you may wish to perform inspections at the subcontractor site; that is, at the company that is sending you the raw materials you use in your product. Whatever works best for you.

Overall Summary of the 1994 Revisions to 4.6 Purchasing

This section is somewhat the same with slight changes throughout, mainly in the definition of particular requirements to clarify what is required. Purchasing procedures must now be documented, which means that some companies weren't adequately doing this before. The revision introduces the issue that subcontractors must be not just selected, but evaluated, and that the subcontractor's quality system will be one of the issues evaluated. So there's a subtle push to buy from other ISO 9000–certified firms. The following changes are critical:

- The type and extent of control over subcontractors must now be defined, as well as the affect of the subcontracted product on the quality of final product. This means some type of document, or change to an existing document, should be developed to address this change.

- The clause about ensuring quality system controls has been eliminated, but only because it is out of place in this section and addressed elsewhere.

- A new section on supplier verification has been added in which if product is verified at subcontractor premises, the supplier will specify how this is to be done in the purchasing documents. This is akin to a reverse contract review requirement, in that the subcontractor must deliver the product as defined in the purchase order; however, in this instance, the organization is responsible for ensuring that it got what it paid for. This means that a procedure for verifying (and documenting the verification of) product at subcontractor sites must be initiated.

In summary, the procedure for assessing subcontractors both initially and during the relationship is expanded. If not already in place, a procedure for source verification must be implemented.

4.7   Control of customer-supplied product

This section deals with use and storage of any components or products supplied by the customer for use in or shipment of the customer's product. In other words, if your organization provides a turnkey service to certain customers in which you store the customer's products or product components at your site, those products or components must be controlled just as any product or component on your site must be controlled.

However, you do not need to inspect the customer's product other than visually, unless you have been contracted to do so. ISO 9000 expects the customer to be responsible for inspections of components and product they purchase and for ensuring that the product they store on your site is what they ordered. As such, the procedures that you need to have for this element center on verification of incoming product (i.e., the product is not obviously damaged and the amount the customer sends you is correct) and for storage of the product according to appropriate measures.

Overall Summary of the 1994 Revisions to
4.7 Control of Customer-Supplied Product

This section has a new title, mainly to "Americanize" it, and puts the focus on the word "customer" as opposed to "purchaser." Same thing, different country usage. There is slight rewording throughout as well, starting with initiating the clause with the words "where appropriate," since many organizations were thrown for a loop trying to interpret how to apply this requirement. The revision puts the emphasis on control of such products, and expands the wording to state that it applies to whatever the customer supplies that the organization uses for whatever purpose. This is not as evil as it seems, since the intent is to address service firms as well as manufacturers. Again, if the clause does not apply, organizations simply say so.

4.8   Product identification and traceability

This section seeks to ensure that all of the product is identified through documents that enable not just the identification of that product but the traceability of it as well. This element is particularly concerned with unit, lot, or batch traceability that permits identification, recall, and replacement of only those products that are nonconforming. For instance, if it is discovered that a bad batch has been produced, then you need to be able to backtrack through the production process to determine where the original error occurred. By testing at certain intervals, you can be assured that acceptable product prior to a particular test does not need to be retested. To establish that you are in control of this requirement, you need to document how you identify product from the raw material stage to the finished product stage. Most organizations apply a unique identification number, such as a bar code or serial number, to individual products or a date or shift code to product batches or lots.

Overall Summary of the 1994 Revisions to 4.8 Product Identification and Traceability

Again, there was minor surgery to this section in the way of rewording. This section is opened up in that the definition for what types of elements are used to identify product is eliminated so that how the product is identified is left up to the organization. It is twice emphasized, however, that there should be documented procedures for this requirement, so it would be wise to closely examine existing procedures to ensure that this requirement is adequately addressed.

4.9   Process control

Process control procedures can be written any way that is appropriate for the organization—just ensure that these procedures provide systematic control of the process to reduce product or process variations and nonconformances. Processes are controlled through work instructions, procedures, standards of acceptability, operating instructions, flowcharts, reference samples, and the like. By documenting your processes, the objective is to ensure that if all your workers walked out, a new crew could perform the tasks based upon the documentation.

The words "and servicing" were added three times in this section, so it's fairly apparent that servicing is considered by the ISO 9000 technical committee to be a product as well as a process. Keep that in mind when you draft your procedures. While this section would normally be thought of as the key component for ensuring that you produce what you said you would, its relative brevity in light of the other elements of this standard highlights the fact that process control includes all the processes that the organization performs. In other words, not just assembly processes, but also design processes, inspection, and test processes, etc. So refer to this element often when reviewing your procedures against the standard.

Overall Summary of the 1994 Revisions to 4.9 Process Control

Because there's a new section on servicing added to the standard, the word "servicing" is beginning to pop up in other areas as well, including here. Servicing is now included as a process under process control, and needs to be appropriately identified, planned, and documented. The following changes also occur:

- There's a word change from the requirement for work instructions to procedures for defining production, installation, and servicing processes. This is simply for consistency since the use of the term "work instruction" was not previously used elsewhere in this standard. (It now only appears in the new Note 7, which states that procedures may reference work instructions.)

- The addition of the words "documented procedures" under process compliance is the second reference of its kind in this revision. It emphasizes the auditor's right to issue a nonconformance if the process actions don't match the process documentation.

- In no uncertain terms, the standard now requires that criteria for workmanship be defined in layperson's terms and that a reference of some kind be used to ensure compliance. This means that the end results must be obvious to all and evident from the start. "No excuses" production, so to speak.

- A new clause zeros in on ensuring that the equipment used for the process works and in such a way that product quality isn't compromised. No more kludgy solutions.

- The title for the section on special processes was eliminated, but the requirement was not, probably because many organizations misinterpreted the requirement. Now the definition for and implementation of a special process (which is any process that cannot be fully verified after manufacture) is clearer, and the requirement that such processes be performed by trained personnel is emphasized. Meeting this requirement can be done through in-process inspections and tests.

- A new sentence states that the requirements for qualifying process operations are to be specified, including equipment and personnel used. This means that if people and machinery need to be prequalified to perform special processes, they should probably be documented as such.

In summary, servicing issues need to be incorporated into existing documentation, either in the form of new procedures or by including them in existing ones, and standards for workmanship need to be defined and documented. Existing procedures for special processes should also be reviewed for compliance to the revised standard.

4.10   Inspection and testing

4.10.1   General

4.10.2   Receiving Inspection and testing

4.10.3   In-process inspection and testing

4.10.4   Final inspection and testing

4.10.5   Inspection and test records

This element is the first of the three inspection and test criteria that ISO 9000 zeros in on. In addition, the 1994 revisions significantly beefed up the requirements, underscoring the importance of this element to the achievement of an ISO 9000 registration. Essentially, you need to define and document your inspection and test procedures. As is apparent, this includes those inspection and test activities that are performed in receiving, in-process, and final inspection. Remember that receiving inspection does not necessarily need to be performed by the Receiving department. This is true for all three criteria; that is, inspections can be performed by any department as long as it is the most appropriate one to perform the inspection.

Three criteria are key: You need to define the frequency of the inspection or test, the test or inspection method, and the accept-reject criteria. And, of course, you need to ensure that the inspections and tests that are performed are recorded and documented appropriately. There needs to be objective evidence that the test or inspection was performed and whether the product passed or failed, as well as approval for release when the product passes the test or inspection.

Most organizations design a simple series of checkpoints in a process and use checklists with this information. The checklist is passed along with the other product documentation so that all tests and inspections applicable to the particular product are documented, and no step is carried out until the previous test or inspection is approved.

Auditors spend a lot of time reviewing these records, so ensure that yours are adequate and complete.

**Overall Summary of the 1994 Revisions
to 4.10 Inspection and Testing**

Several elements were expanded upon in this section, including adding a general introduction segment that establishes the requirement to document this area and that this information should be in the quality plan or procedures. No longer is an organization required to just do it, but to determine what needs to be inspected and tested, what those activities are, and the accompanying records. Other specific requirements were added:

- Under receiving inspection and testing, a Note has now become a clause that requires the organization to determine not only the amount and nature of receiving inspection, but to determine the amount of control that should be exercised at the subcontractor's premises as well as during receiving inspection. Records were again emphasized. This means that more attention will be given to receiving inspections in subsequent audits and that such inspections must be recorded as defined in the documentation.

- The clause that allows release of incoming product for urgent production purposes clarifies that this release is prior to verification. It doesn't state that this is the only time incoming product can be released, but it implies that no other time is acceptable.

- Under in-process inspection and testing, two clauses were eliminated; one regarding the requirement to identify nonconforming product, the other requiring the establishment of whether the product conforms to specified requirements. This is not because they are no longer required, but because they are addressed elsewhere in the standard.

- No major changes under final inspection and testing.

- Under inspection and test records, the verbiage is expanded from the requirement for the product to pass inspection to state that the product has to have been inspected and that records must state whether it passed or failed. It goes on to state that if the product fails to pass, then the procedures for handling nonconforming product kick in.

- There's a final requirement again for records to be kept. That's three references to records in a single element—which means that the auditors will be looking at this area closely.

In summary, this area must be thoroughly documented and records maintained. It will be an issue during external surveillance audits.

4.11   Control of inspection, measuring, and test equipment

4.11.1   General

ISO 9000 separates the equipment from the process; in this case, the inspection, measuring, and test equipment is separated from the test and inspection processes. This separation underscores the importance of ensuring that all equipment used for these purposes is controlled and functioning correctly. What needs to be done is to take a thorough inventory of all such equipment on your premises and to ensure that the equipment is identified, calibrated, and that records of the calibration are maintained. Such equipment includes all software used to ensure process or product accuracy. If the equipment is self-calibrating, then you need to provide proof that the calibration being performed by the equipment is within specifications. Get the equipment manufacturer to provide guidelines and assistance on this issue.

Most organizations produce a matrix that identifies the test, measuring, and inspection equipment. The matrix or list should include the equipment identification, location, calibration frequency, calibration method, applicable tolerances, and who is responsible for performing the calibration. The equipment itself needs to be tagged with a sticker that indicates when it was last calibrated and when calibration is due. Even when this information is also included on the calibration matrix, it must also be attached to the equipment itself. It's also a good idea to check all such equipment before an audit to ensure that nothing has been forgotten. The one piece of equipment that slips past your notice is sure to be the one out of calibration when the auditor arrives.

Remember that not everything needs to be calibrated if the measurements aren't critical. But if the equipment requires calibration, make sure it is calibrated to a reference standard. National standards are best for this purpose.

4.11.2 Control procedure

It is very important that this process of checking test, measuring, and inspection equipment be controlled, so important that ISO 9000 breaks it out as a separate clause within the element. Note that the element begins with determining what measurements need to be made; you may not be taking enough measurements during your processes to ensure that the product meets your design or customer criteria. Or, you may be using inappropriate equipment. Thus a review is necessary, as well as some form of identification of what needs to be checked. The matrix mentioned above will satisfy the second criteria in Sec. b. You will also need to specify how you will inspect the equipment separate from using the equipment manufacturer's documentation. In other words, you can't just say you use their test manual because their test manual won't define what *you* should do if the equipment doesn't pass the inspection. Your procedures will specify how you identify the equipment as being out of calibration and how you prevent the equipment from being used until it can be calibrated again. When drafting your procedure for this process, take a good look at all nine points in this clause to ensure that each is satisfied.

Overall Summary of the 1994 Revisions to 4.11 Control of Inspection, Measuring, and Test Equipment

This section has been expanded so that test equipment is now defined as comprising test software as well as test hardware. It takes out the clause about who owns the equipment mainly because it's unacceptable to disclaim responsibility for a piece of equipment that is being leased as opposed to one that is owned. If the organization is using the equipment, it is responsible for the equipment's functionality, period.

This element, already significant, has been expanded and broken into subsections for clarification. The following changes are noted:

- The general introduction rephrases the responsibility for maintaining this equipment and access to the related data by the customer. No big changes, just clarification.

- A new subsection on control procedures accommodates previous points, but expands on the fact that all equipment which can affect product quality must be identified and calibrated or adjusted at prescribed intervals (not just when needed). This means a list must be developed and a schedule created, then adhered to.

- A new clause in the control procedure requires that the process used to calibrate equipment must be defined, as well as specific checks and acceptance criteria, and what needs to happen when the results are out of spec. This means that even if the person performing calibration is using a manual, and if the manual doesn't specify the calibration parameters, a documented procedure must.

- The requirements for what needs to be included in the documented calibration procedures were removed; this is now left up to the organization to define.

- Note 18 was added to refer to ISO 10012 for guidance in this area. This means the auditors will closely follow any recommendations in that standard.

In summary, the changes to this section now require addressing relevant test software in more detail; documenting what gets calibrated and when as well as the process for performing these steps; and that the organization needs to refer to ISO 10012 to ensure that what's suggested in that standard has been incorporated in this one. Not so much new requirements as a more in-depth look at existing procedures. Best to do a thorough review of what's in place to ensure full compliance.

4.12 Inspection and test status

This final element covering inspection and test criteria focuses on the need to ensure conformity of the product that is being tested and inspected. By emphasizing the need to define the tests and inspections in the documented quality plan and/or documented procedures, the standard once again emphasizes the need to ensure that if the process is correct, the end result will be as well. Again, scrutinize your processes and documentation to ensure that the checklists you use and the procedures you have written address the criteria in these three test and inspection elements.

Keep in mind that even if you seldom have problems or nonconformances, you still need to have a procedure for how you handle rejected product. You can stamp a batch card, attach Quality Hold stickers or tags, or quarantine product in a specific location. Just remember to identify and segregate it in some way to ensure that only accepted product is released.

**Overall Summary of the 1994 Revisions
to 4.12 Inspection and Test Status**

This element was loosened up a bit in that the recommended means by which inspection and test status of a product are identified were eliminated, and how this is done is now left up to the individual organization. However, a new statement has been added which requires that how status is identified must be defined in the quality plan and/or documented procedures. So, they no longer care how you do it, just that you document how you're going to do it.

4.13   Control of nonconforming product

4.13.1   General

4.13.2   Review and disposition of nonconforming product

One of the key issues in any quality system is to ensure that product which is out-of-spec or doesn't meet acceptance criteria, that is, nonconforming product, is not passed along to the customer. This element applies to all stages of manufacturing and processing from receipt of raw materials to finished product. As the element points out, such nonconforming product must be identified, segregated, and evaluated and the results documented. Once the nonconforming product has been reviewed, there are essentially four choices regarding its disposition:

1. You can rework or fix it to meet specifications.

2. You can accept it as is by concession.

3. You can regrade it to a lower designation.

4. You can reject and scrap it.

Whatever you do, you must document your decision as well as the process for arriving at that decision. Remember to specify who can release nonconforming product and how you would recall a product after release.

Many companies elect to develop a Quality Hold procedure of some sort. Thus when product or equipment is found to be unacceptable, the person who discovers the nonconformity places a sticker or tag or label on the product or equipment stating that it is not to be used. If the product is discovered during assembly, the entire line may be placed on hold while a supervisor is located and notified for resolution. One of those two persons usually writes an Incident Report stating the problem and what was done to resolve it. Such Incident Reports are then forwarded to the quality teams or committees who review nonconformances for short-term and long-term corrective and preventive actions. The product or equipment can then be released only after the appropriate approval authority has signed off the Incident Report or other similar documentation.

In addition, define what you do if product that does not meet specifications needs to be released in order to meet time or production requirements. Do you need to obtain a concession from the customer or just identify the batch or product as such? Also, don't forget to reinspect product that has been reworked.

Overall Summary of the 1994 Revisions to 4.13 Control of Nonconforming Product

Very little was changed in this section other than a few words for clarification's sake. The procedures must now be documented, however, if they were not previously.

4.14   Corrective and preventive action

4.14.1   General

4.14.2   Corrective action

4.14.3   Preventive action

Finding nonconformances is one thing, correcting the process to ensure that the problem doesn't occur again is the next step. Corrective and preventive actions are treated as two separate issues. First, when a problem is found it needs to be corrected immediately in order to keep production going. Short-term fixes, whether done on the spot or shortly thereafter, are known as corrective actions. However, if a problem continues to recur, a long-term solution may be necessary. Such a solution is known as preventive action to ensure that the problem doesn't occur again. Some preventive actions are developed by reviewing related problems. If the same problem continually recurs, the fault may lie with the process. Using Incident Reports to zero in on similar nonconformances can help ensure that such nonconformances are eliminated from the organization. This clause seeks to ensure that the review of the problems as well as a review of the solutions to confirm the validity and effectiveness of the fix are performed.

You may choose to create quality teams independent of the quality department within each department. Those department quality teams can then address the problems they run up against and propose solutions. It is often said that the ones who best know how to fix the problem are the ones who are performing the process. Take advantage of all the people in your organization.

Keep in mind that these quality teams must also keep minutes and that the results of the solutions the quality teams propose should be reviewed by management during the management meetings required in 4.1.3.

Corrective and preventive action are the heart of any quality system and can help eliminate problems and reduce operating costs. In order to address corrective and preventive actions effectively, defects and nonconformances should be classified to establish the type of defects, and then quantify the frequency with which that particular defect occurs. In addition, how you handle customer complaints is addressed by this section.

Overall Summary of the 1994 Revisions to 4.14 Corrective and Preventive Action

This section was greatly expanded and broken into subsections, most notably through the separation of corrective action from preventive action and the corresponding activities for both. The following requirements were added:

- A general introduction section states that documented procedures for both corrective and preventive action are now required. It also states that the action taken in response to a corrective/preventive action should be to a degree appropriate to the magnitude of the issue. So if the problem is statistically irrelevant, or would cause more problems to implement than are justified, it's OK not to go overboard just to make an auditor happy.

- If the organization is not already doing so, recording changes to the procedure that are the result of a corrective/preventive action is now required.

- A new subsection on corrective action highlights customer complaints and reports of product nonconformities to be handled effectively. Unfortunately, just what effective handling is has not been defined.

- The corrective action subsection also states that investigations into the cause of nonconformities shall include not just product, but process and quality system issues as well, and that any investigation must be recorded. On a practical level, this means that any forms used to investigate product nonconformities should have sections added to cover process and quality system aspects too.

- The corrective action subsection also states that the corrective action which needs to be taken must be determined. No loose ends allowed.

- The preventive action subsection requires using a range of information sources to detect, analyze, and eliminate potential causes of nonconformances. This means the auditors will be looking for an in-depth, thorough analysis of problems, not just a surface-level solution.

- The preventive action subsection also requires that the steps to deal with problems requiring preventive action need to be determined and that controls need to be put in place to ensure effectiveness. Auditors will look to ensure that not only were the action steps defined but that there were controls as well. This will need to be on a document or record somewhere as proof of compliance with this element.

- Finally, the preventive action subsection requires information regarding preventive action be submitted to upper management for review. While no management action is directly required, it implies that management should act upon it in some way. This means that management needs to be conversant in preventive action issues, and that these issues are discussed during the quality system reviews.

In summary, this is the continuous improvement requirement that everyone has been expecting, although why it isn't called continuous improvement is a mystery. What compliance to these changes will require is a review of the existing corrective and preventive action process with an eye to expanding upon any forms, documented procedures, work instructions, etc., to break out the two. It's also a good idea to include them in management reviews as separate agenda items.

4.15   Handling, storage, packaging, preservation, and delivery

4.15.1   General

4.15.2   Handling

4.15.3   Storage

4.15.4   Packaging

4.15.5   Preservation

4.15.6   Delivery

This clause addresses entire processes that are performed by specific departments as well as steps within the processes performed by just about everyone. Handling, for instance, refers not just to receipt of product but that any time the product is handled during assembly or production, it is to be handled according to procedures. The storage clause is appropriate to warehouse functions, but can also be applied to other stockrooms that may contain items such as product labels. Cycle counts and inventory inspections need to occur at regular intervals, and the receipt and dispatch procedures of product from one area to another need to be defined. The most important point, however, is to prevent product degradation or damage during either storage or transit.

What you'll probably find is that your current receiving, inventory, warehouse, and shipping procedures are adequate for ISO 9000 and may only need a few checklists to meet the documentation requirements. One thing to look for—there are often weight scales in receiving and shipping that are overlooked because most companies check their production department for calibration equipment. The equipment in receiving and shipping also falls under the inspection, measuring, and test equipment element (4.11).

Overall Summary of the 1994 Revisions to 4.15 Handling, Storage, Packaging, Preservation, and Delivery

The most significant change in this element is the addition of "preservation" to the list of required items. Again, the procedures must be documented if they are not already. The following changes also apply:

- Under storage, the word "secure" was eliminated as applies to storage and replaced with "designated." This is a fine-line differentiator but now allows the organization to store product where appropriate, thereby acknowledging that security or locking up product is not always appropriate.

- Under packaging, the word "packaging" was added so that packing and packaging are both to be controlled; again, a fine-line differentiator, but these are two different words and while the original intent was to include both within the scope of the standard, obviously someone argued the point.

- A new section on preservation has been added that now breaks out a previous requirement to preserve and segregate product into a separate category. While this is a new section, it does not impose new requirements.

In summary, little has changed with this element beyond semantics. There are no new requirements to implement.

4.16 Control of quality records

Quality records are distinct from document and data control (4.5) in that these records are the output of the processes and provide a history of the event. Quality records are the design documentation, training records, calibration reports, and internal audit reports. Quality records are the receiving and shipping documents, and inspection and test reports. Quality records are the customer orders and the paperwork that follows the order throughout the company. What you need to do is define where and how long each quality record is to be maintained, and who is responsible for managing them.

One method is to create a matrix that lists the various types of quality records used by your company and specify how long each of those records is to be kept. Usually your documented processes will end with a statement to the effect that the document is filed by a certain department or forwarded to another. If you have it forwarded to another, you need to then ensure that the department that receives the forwarded document or record either files it or forwards it again, until the document or record is filed. Sometimes all documentation returns to the customer support department for filing in the customer files. Sometimes the documentation ends up in Inventory. Each company is a little different. You already have a resting place for most documents. This element provides you with the methodology to ensure that all appropriate documents are maintained and that you know where those documents are.

When there are too many documents to be kept in the active files, an archive solution may be determined. If you elect to store last year's documents or records in a warehouse, you need to state where they can be found.

Many organizations check with their legal departments to ensure that documents which are required by law to be stored for a certain period of time are kept for that period of time. Such legal requirements fall under the statutory and regulatory requirements to which ISO 9000 is now paying heed.

Overall Summary of the 1994 Revisions to 4.16 Control of Quality Records

Little has changed in this section beyond a requirement that the procedure be documented and include a reference to access issues; that is, who shall have access to quality records. The records must now "demonstrate conformance to requirements" as opposed to "achievement of required quality," which is more or less the same thing. And the prevention of record deterioration or damage as opposed to minimizing this occurrence (apparently they expect these records to withstand the test of time ad infinitum) is now required. Even so, given that the quality record procedure may need to be subtly changed to address access issues, no major changes are required to be in compliance with this element.

4.17   Internal quality audits

ISO 9000 relies not only on external third-party auditors from the registrar to check the effectiveness of your quality system, but requires that you perform internal quality audits using your own personnel as well. The external auditors will also check that such internal quality audits have been performed and that the internal quality audit function has also been audited.

There are a number of different solutions to addressing this requirement. Some organizations designate a small team of internal quality auditors who may also be product inspectors. This dedicated team performs all the internal audits. If your company doesn't have the extra available personnel, or prefers not to hire additional help, another solution is to create large teams composed of people from every department. The advantage to this solution is that more people are familiar with the auditing process and will better understand how to be audited as well as the best way to audit. On the other hand, this solution adds to the job responsibilities and duties of employees who may already be stretched to the limit. Adding auditing tasks may not be looked on favorably, and you will need to ensure that each auditor is trained and retrained appropriately.

While the standard does not specify how often the internal audits need to occur, a good rule of thumb is at least once a quarter with the goal of auditing the entire system over the course of a year. Alternatively, you can use the period immediately before an external audit to perform an internal audit and prep the organization for the upcoming external surveillance audit. Or, you can do both.

For an internal audit, you will need to have an audit schedule that lists the departments or functions that need to be audited, when they will be audited, and whether the audit was performed on time or rescheduled. Then use checklists to ensure that the audit is complete. Finally, don't forget to follow up on audit findings and nonconformances to ensure resolution.

At a minimum, each department or function should be audited at least once a year. Internal audit results that highlight nonconformances need to be treated in the same way that an external audit result is addressed: you should fix the problem on the spot or within the month, and provide an explanation of the proposed corrective or preventive action solution. Such internal audit results should also be discussed during the appropriate quality and management review meetings.

Overall Summary of the 1994 Revisions to 4.17 Internal Quality Audits

This element has been expanded slightly but not much. Essentially, documented procedures are now required for the planning and implementation of internal audits, and the related results must comply with both these arrangements and the effectiveness of the overall system.

It's now in black and white that audit personnel must be independent of the function being audited. Any company that skirted this issue in the past will have to come up with another solution.

Rather than just state that audits and follow-up activities must be in accordance with audited procedures, the standard now states that follow-up activities must verify and record the implementation and effectiveness of the corrective action taken. This ties internal quality audits much closer to the corrective action procedure, as it should be. It also directly states that audit results should be reviewed during management reviews and that additional guidance can be obtained from 10011-1, 10011-2, and 10011-3. That's the kicker. While most organizations are probably already meeting the additional requirements, it would be wise to review the documented procedures against these other standards to ensure compliance.

4.18   Training

Simply put, everyone needs to be trained to do his or her job, and proof of that training needs to exist. Therefore, a training folder should be created for each employee which contains copies of the classes and training that employees have received which makes them eligible to perform their job. You will need to provide proof of that training in the form of certificates or attendance records or some such appropriate document.

In some cases, you may need to create a matrix of the training required for certain positions. For instance, if you have numerous assembly or production workers who require training on a semiannual basis, you might want to create a matrix of the training required for each job function, the names of the persons who need to be trained, the dates of the most recent training, and the dates scheduled for retraining. Many times, training can be accomplished in a group classroom situation. At the end of the class, the names of the persons who attended are recorded on an attendance checklist and copied to their training folder.

Don't forget that if you provide internal training using internal instructors, those instructors must also be qualified to perform such training. Proof of their qualification must be included in the instructor's training folder. It's also a good idea to make ISO quality system training part of new employee orientation training to ensure awareness of all employees concerning the standard's requirements.

Overall Summary of the 1994 Revisions to 4.18 Training

Only one word was added to this section, and that is a requirement for the training procedures to be documented if they are not already.

4.19   Servicing

This clause separates service activities from other processes. Many companies do not perform this function and can simply address this requirement by stating, "Our company does not perform after-sales servicing of the products we sell. If and when such activities are included in the business functions, these activities will be addressed by this quality plan and system." That's all you need to say to satisfy an auditor.

If your organization does perform servicing, you need to ensure that all terms and conditions are defined and then develop processes for fulfilling those terms and conditions. Develop documentation that states what the servicing processes are, much like the rest of the processes that have been included in your quality procedures. The same issues apply to servicing as to all other processes; the persons performing the servicing must be trained; the service records must be maintained; the service procedure must be documented.

You might want to take advantage of this to formalize an escalation procedure for handling difficult service issues in the event that a product cannot be fixed within a guaranteed time frame, etc. Ensuring that your customers know of the escalation procedure provides them with added assurance that your organization is dedicated to providing a service product of the highest quality.

Overall Summary of the 1994 Revisions to 4.19 Servicing

A new section on servicing was added to ISO 9002/9003, again pulled in from ISO 9001. It consists of one sentence that requires the organization to address servicing issues where servicing is performed by the organization. At this point in time it simply requires that procedures for performing, verifying, and reporting servicing be put in place without going into any detail on how this should be done. This means that if a service department exists at the site, it must be incorporated into the quality system for the site to maintain its registration.

4.20   Statistical techniques

4.20.1   Identification of need

4.20.2   Procedures

The requirement for statistical techniques is the closest ISO 9000 comes to the old-fashioned quality control practices as opposed to the more modern quality assurance activities. Quality control tended to focus more on the numerical aspects of the process; that is, the measurements for ensuing that defects were kept to a certain level and that quality goals were measured and reached. Whether using a tried and true technique or a new approach, whenever less than 100 percent of a product is inspected or tested, some kind of sampling system must be used to ensure that product conforms to an acceptable quality level (AQL). Routine inspections or tests at fixed time intervals also satisfy this requirement. The key is to identify suitable techniques that enable your organization to verify conformance of products and processes.

As is apparent, ISO 9000 places no specific requirements in this area, just that the organization perform some sort of measurement technique for gathering statistics that provide insight into the effectiveness of the various processes. Which processes are measured is left entirely up to the organization. Therefore, identify what you want to track and develop a procedure for tracking that information. If you wish to get more mileage out of the statistics that you track, you can discuss such information at quality and management meetings, but this is not required.

Overall Summary of the 1994 Revisions to 4.20 Statistical Techniques

Interestingly enough, while this element has not been enlarged upon very much, it has been broken out into two subsections. A precursor of things to come? The new subsections require that the need for statistical techniques be identified and documented procedures established to address this element. Identification of the need does not, however, require a corresponding documentation of that identification. And while the words "where appropriate" were eliminated—meaning that all ISO-certified organizations must employ some kind of statistical technique—the requirement for the statistical techniques to verify the acceptability of process capability and product characteristics was eliminated. This means that the organization can measure whatever it likes based upon what has been idcntified. Go figure. Essentially, unless the organization was not performing any statistical techniques previously, no additional steps are required for compliance.

Elements Requiring Documented Procedures

4.1   Management responsibility—Organizational responsibility should be documented (4.1.2.1).

4.2   Quality system—The quality system should have already been documented through a quality manual (4.2.1), and quality planning should be documented (4.2.3).

4.3   Contract review—Contract review procedures should now be documented (4.3.1).

4.4   Design control—Procedures should be established, maintained, and documented (4.4.1).

4.5   Document control—Document control procedures should be documented (4.5.1).

4.6   Purchasing—Purchasing procedures should now be documented (4.6.1).

4.7   Control of customer-supplied product—Where appropriate, these procedures should be documented (4.7.1).

4.8   Product traceability and identification—Where appropriate, these procedures should be documented (4.8.1).

4.9   Process control—There procedures should have already been documented (4.9.1).

4.10   Inspection and testing—These procedures should now be documented (4.10.1).

4.11   Inspection, measuring, and test equipment—These procedures should now be documented (4.11.1).

4.12   Inspection and test status—This information should be documented in either the quality plan and/or documented procedures.

4.13   Control of nonconforming product—This procedure should now be documented (4.13.1).

4.14   Corrective and preventive action—These procedures should have already been documented (4.14.1).

4.15   Handling, storage, packaging, preservation, and delivery—These procedures should have already been documented (4.15.1).

4.16   Quality records—These procedures should now be documented.

4.17   Internal quality audits—These procedures should now be documented.

4.18   Training—These procedures should now be documented.

4.19   Servicing—These procedures should be documented.

4.20   Statistical techniques—These procedures should now be documented (4.20.2).

Essentially, everything should now be documented.

B

Sample Quality
Policy Manual Template

This appendix contains a sample Quality Policy Manual template that you can use as a starting point to create your own Quality Policy Manual. All the text in brackets is provided to give you pointers on what should be included in that section of the manual. If you replace the text in brackets with answers to the questions and responses to the suggestions, you will have the first draft of your own Quality Policy Manual. The headers at the top of the pages use the abbreviation QPM to stand for Quality Policy Manual. As with all your quality documentation, each page needs to include in the header:

- The title of the document
- Who it belongs to (your company name)
- The issue or revision number
- The numbering system for the document (in this case QPM 001, which stands for Quality Policy Manual, section number 001)
- The page number
- The date the document is effective (i.e., the latest update)

This format works. It has worked for numerous companies and is what an external auditor would expect to see. Good luck and Bon ISO!

[Your Company Name]

ISO 900X

Quality Policy Manual

QUALITY POLICY		QPM 001
Issue: 1.0		Page 1 of 1
Your Company Name	Amendments and Issues	Eff. Date: D/M/Y

QPM 001:   AMENDMENTS AND ISSUES

ISSUE	AFFECTED PAGES	DESCRIPTION	EFFECTIVE DATE
1.0	All	First issue of new manual	Day, Month, Year (D,M,Y)

QUALITY POLICY		QPM 002
Issue: 1.0		Page 1 of 1
Your Company Name	Approvals	Eff. Date: D/M/Y

QPM 002:   APPROVALS

_____ President

_____ Vice President

_____ Sales Manager

_____ Financial Controller

_____ Human Resources Manager

_____ Manufacturing Manager

_____ Quality Manager

_____

_____ ISO 900X Project Leader

[The names and signatures of the persons who will sign your top-level quality policy manual need to be provided at the beginning of this document similar to this layout.]

QUALITY POLICY		QPM 003
Issue: 1.0		Page 1 of 1
Your Company Name	Table of Contents	Eff. Date: D/M/Y

QPM 003:   **TABLE OF CONTENTS**

[The Procedure Number listed above corresponds to the element number in the ISO 9000 standards. The Procedure Title reflects that element. The Procedure Length refers to page 1 of 1, or however many pages it takes to address that particular element.]

QUALITY POLICY   QPM 004
Issue: 1.0   Page 1 of 1
Your Company Name   Company Structure and Business Statement   Eff. Date: D/M/Y

QPM 004:   COMPANY STRUCTURE AND BUSINESS STATEMENT

1.0   Company Structure

[Provide a brief description about the company. Who are you, what do you do, what industries do you sell to, and where do you sell?]

2.0   Business Statement

[What do you make, and what is it used for? Briefly describe your business goals.]

QUALITY POLICY		QPM 010
Issue: 1.0		Page 1 of 1
Your Company Name	Company Quality Policy	Eff. Date: D/M/Y

QPM 010:   COMPANY QUALITY POLICY

[How would you express the long-term goals and operating philosophy of your company in a way that addresses overall corporate and product quality? Encompass why quality is important to your company and how such an approach is beneficial.]

[Picture this as a short, three- to four-sentence statement that you can post in the facility as well and distribute to employees if needed.]

QUALITY POLICY		QPM 011
Issue: 1.0		Page 1 of 1
Your Company Name	Company Organization Chart	Eff. Date: D/M/Y

QPM 011:   COMPANY ORGANIZATION CHART

[Provide a copy of the company's organization chart. Keep this general by focusing on departments or functions. Do not provide personnel names, just job titles at the most.]

QUALITY POLICY		QPM 012
Issue: 1.0		Page 1 of 1
Your Company Name	Quality System Responsibilities	Eff. Date: D/M/Y

QPM 012:   QUALITY SYSTEM RESPONSIBILITIES

1.0   President

2.0   Vice President

3.0   ISO 900X Project Leader

4.0   etc.

[List the titles of all personnel, starting with upper management, who participate in quality activities and their role in the ISO 900X quality system. You can group lower-level job titles under one section when those jobs have lesser quality responsibilities. Remember that everyone involved in product and process quality has quality system responsibilities.]

QUALITY POLICY		QPM 013
Issue: 1.0		Page 1 of 1
Your Company Name	Management Review	Eff. Date: D/M/Y

QPM 013:   MANAGEMENT REVIEW

1.0   Policy

[How will management review the quality system? When will they do it (frequency), who will conduct it, and what specific elements will they review?]

2.0   Responsibilities

[Who will be responsible for what at the review, and who will attend?]

3.0   Documentation

[What procedures do you use that support management review activities?]

QUALITY POLICY		QPM 020
Issue: 1.0		Page 1 of 1
Your Company Name	Quality System	Eff. Date: D/M/Y

QPM 020:   QUALITY SYSTEM

1.0   Policy

[What requirements are your quality system intended to meet, how will you ensure that these requirements are met, and what other quality systems are in place that will enhance or affect this system?]

2.0   Responsibilities

[Who is responsible for ensuring that this policy is enforced?]

3.0   Documentation

[What procedures do you use that support quality system requirements?]

QUALITY POLICY   QPM 030
Issue: 1.0   Page 1 of 1
Your Company Name   Contract Review   Eff. Date: D/M/Y

QPM 030:   CONTRACT REVIEW

1.0   Policy

[How are contracts and proposals handled? How is the order entry process performed? What do you look for, and what determines when a contract can be
signed?]

2.0   Responsibilities

[Who is responsible for ensuring that the above policy is fulfilled?]

3.0   Documentation

[What procedures do you use that support contract review and order processing
activities?]

QUALITY POLICY   QPM 040
Issue: 1.0   Page 1 of 1
Your Company Name   Design Control   Eff. Date: D/M/Y

QPM 040:   DESIGN CONTROL

1.0   Policy

[What is your policy regarding design control? What do you do to ensure the integrity of the product and the process? What are your goals?]

2.0   Responsibilities

[Who is responsible for ensuring that the above policy is fulfilled?]

3.0   Documentation

[What procedures do you use that support design control activities?]

QUALITY POLICY   QPM 050
Issue: 1.0   Page 1 of 1
Your Company Name   Document Control   Eff. Date: D/M/Y

QPM 050:   DOCUMENT CONTROL

1.0   Policy

[How are documents handled, specifically the quality manual documentation? What do you ensure regarding documentation control and handling? What are your goals?]

2.0   Responsibilities

[Who is responsible for ensuring that the above policy is fulfilled?]

3.0   Documentation

[What procedures do you use that support document control activities?]

QUALITY POLICY		QPM 060
Issue: 1.0		Page 1 of 1
Your Company Name	Purchasing	Eff. Date: D/M/Y

QPM 060:   PURCHASING

1.0   Policy

[How is purchasing handled? What do you do to ensure the provision and delivery of quality materials? How are vendor monitoring and assessment part of this policy? What are your goals?]

2.0   Responsibilities

[Who is responsible for ensuring that the above policy is fulfilled?]

3.0   Documentation

[What procedures do you use that support purchasing activities?]

QPM 070:   CONTROL OF CUSTOMER-SUPPLIED PRODUCT

1.0   Policy

[How is the product supplied by your customers handled? What do you do to ensure procedures for verification, storage, and maintenance of customer material? What are your goals?]

2.0   Responsibilities

[Who is responsible for ensuring that the above policy is fulfilled?]

3.0   Documentation

[What procedures do you use that support and control customer-supplied product activities?]

QUALITY POLICY		QPM 080
Issue: 1.0		Page 1 of 1
Your Company Name	Product Identification and Traceability	Eff. Date: D/M/Y

QPM 080:   PRODUCT IDENTIFICATION AND TRACEABILITY

1.0   Policy

[How is product identification and traceability handled to ensure that all materials and products are clearly identified throughout the company, from receipt to production to distribution? What do you do to ensure traceability? What are your goals?]

2.0   Responsibilities

[Who is responsible for ensuring that the above policy is fulfilled?]

3.0   Documentation

[What procedures do you use that support product identification and traceability requirements?]

QPM 090:   PROCESS CONTROL

1.0   Policy

[How are processes that affect, and that are affected by, quality controlled? What do you do to ensure process monitoring and control? What are your goals?]

2.0   Responsibilities

[Who is responsible for ensuring that the above policy is fulfilled?]

3.0   Documentation

[What procedures do you use that support process control activities?]

QUALITY POLICY		QPM 100
Issue: 1.0		Page 1 of 1
Your Company Name	Inspection and Testing	Eff. Date: D/M/Y

QPM 100:   INSPECTION AND TESTING

1.0   Policy

[What procedures and controls for product verification are in place for inspection and testing? Receiving inspection? In-process testing? Final test and inspection? Inspection and test records? What are your goals?]

2.0   Responsibilities

[Who is responsible for ensuring that the above policy is fulfilled?]

3.0   Documentation

[What procedures do you use that support inspection and testing activities?]

QPM 110:   CONTROL OF INSPECTION, MEASURING, AND TEST EQUIPMENT

1.0   Policy

[How are inspection, measuring, and test equipment controlled, calibrated, and maintained? What are your goals?]

2.0   Responsibilities

[Who is responsible for ensuring that the above policy is fulfilled?]

3.0   Documentation

[What procedures do you use that control and identify inspection, measuring, and test equipment?]

QUALITY POLICY		QPM 120
Issue: 1.0		Page 1 of 1
Your Company Name	Inspection and Test Status	Eff. Date: D/M/Y

QPM 120:   INSPECTION AND TEST STATUS

1.0   Policy

[How is inspection and test status of the product ensured? What do you do to verify and record this status? What are your goals?]

2.0   Responsibilities

[Who is responsible for ensuring that the above policy is fulfilled?]

3.0   Documentation

[What procedures do you use that support and identify inspection and test status activities?]

QPM 130:   CONTROL OF NONCONFORMING PRODUCT

1.0   Policy

[How do you ensure that nonconforming product is clearly identified and segregated? What evaluation and disposal controls are in place? What are your goals?]

2.0   Responsibilities

[Who is responsible for ensuring that the above policy is fulfilled?]

3.0   Documentation

[What procedures do you use to control, identify, and segregate nonconforming products?]

QUALITY POLICY		QPM 140
Issue: 1.0		Page 1 of 1
Your Company Name	Corrective and Preventive Action	Eff. Date: D/M/Y

QPM 140:   CORRECTIVE AND PREVENTIVE ACTION

1.0   Policy

[How is corrective action handled? Is there a corrective action team? What do you do to ensure that nonconforming products are analyzed, recorded, and the potential causes eliminated? What are your goals?]

2.0   Responsibilities

[Who is responsible for ensuring that the above policy is fulfilled?]

3.0   Documentation

[What procedures do you use that support corrective and preventive action processes?]

QUALITY POLICY		QPM 150
Issue: 1.0	Handling, Storage, Packaging,	Page 1 of 1
Your Company Name	Preservation, and Delivery	Eff. Date: D/M/Y

QPM 150:   HANDLING, STORAGE, PACKAGING, PRESERVATION, AND DELIVERY

1.0   Policy

[How is the product—and all materials that make up the product or affect it in any way—handled, stored, packaged, preserved, and delivered? What controls exist? What are your goals?]

2.0   Responsibilities

[Who is responsible for ensuring that the above policy is fulfilled?]

3.0   Documentation

[What procedures do you use that support handling, storage, packaging, preservation, and delivery activities?]

QPM 160:   CONTROL OF QUALITY RECORDS

1.0   Policy

[Are quality records maintained in a way that demonstrates achievement of ISO quality standards? Do they support the effective operation of the quality system? What are your goals?]

2.0   Responsibilities

[Who is responsible for ensuring that the above policy is fulfilled?]

3.0   Documentation

[What procedures do you use that support and control quality records?]

QPM 170:   INTERNAL QUALITY AUDITS

1.0   Policy

[How are internal quality audits conducted? How often are they conducted? Who conducts them, and how are the results reported? What other quality audits are performed? What are your goals?]

2.0   Responsibilities

[Who is responsible for ensuring that the above policy is fulfilled?]

3.0   Documentation

[What procedures do you use that support internal quality audit activities?]

QUALITY POLICY		QPM 180
Issue: 1.0		Page 1 of 1
Your Company Name	Training	Eff. Date: D/M/Y

QPM 180:   TRAINING

1.0   Policy

[What do you do regarding the selection, development, and training of employees through job descriptions, annual appraisals, succession plans, and education/training programs? Is there an annual training/development plan? What are your goals?]

2.0   Responsibilities

[Who is responsible for ensuring that the above policy is fulfilled?]

3.0   Documentation

[What procedures do you use that support training requirements and activities?]

QUALITY POLICY		QPM 190
Issue: 1.0		Page 1 of 1
Your Company Name	Servicing	Eff. Date: D/M/Y

QPM 190:   SERVICING

1.0   Policy

[How do you define terms and conditions for service contracts? What services do you provide and where do you maintain the service records—at your site or the customer's site? Do you have an escalation procedure when problems arise? What are your goals?]

2.0   Responsibilities

[Who is responsible for ensuring that the above policy is fulfilled?]

3.0   Documentation

[What procedures do you use that support servicing activities?]

QUALITY POLICY
Issue: 1.0
Your Company Name   Statistical Techniques

QPM 200
Page 1 of 1
Eff. Date: D/M/Y

QPM 200:   STATISTICAL TECHNIQUES

1.0   Policy

[Do you utilize statistical techniques? When and where? Upon what criteria do you base your statistical techniques, and how are statistical techniques used by your company? What are your goals?]

2.0   Responsibilities

[Who is responsible for ensuring that the above policy is fulfilled?]

3.0   Documentation

[What procedures do you use that support statistical technique activities?]

C

Sample
Nonconformance Report

NONCONFORMANCE REPORT

NONCONFORMANCE
REPORT NO:

COMPANY NAME:

DEPARTMENT/AREA AUDITED: DATE OF AUDIT:

DEPARTMENT REPRESENTATIVE: AUDITOR:

NONCONFORMANCE: MAJOR/MINOR

CLAUSE 4.____OF ISO 900____

CAUSE IDENTIFICATION/CORRECTIVE ACTION PROPOSED:

AGREED TIME FOR IMPLEMENTATION: RESP. FOR ACTION:

DEPARTMENT REP'S SIGNATURE:

AUDITOR'S SIGNATURE: DATE:

CORRECTIVE ACTION COMPLETED
 SATISFACTORILY: YES/NO

ANY COMMENTS:

SIGNATURE: DATE:

Sample
Quality Assessment
Report

QUALITY ASSESSMENT REPORT

QUALITY ASSESSMENT REPORT NO:

COMPANY NAME:

NAME OF ASSESSOR(S):

DEPARTMENT/AREA AUDITED: DATE OF AUDIT:

DOCUMENT ASSESSMENT CONDUCTED
AGAINST: ISO 900___

DOCUMENTED QUALITY SYSTEM REFERENCE:

NAMES OF PERSONNEL SEEN: JOB TITLE:

PURPOSE OF ASSESSMENT:

CONCLUSION:

LEAD ASSESSOR SIGNATURE: DATE:

COMPANY REPRESENTATIVE SIGNATURE: DATE:

E

Quality System Registrars

This is a short list of registrars based in the United States and Canada. Please contact the Registrar Accreditation Board (RAB) for a more complete listing and to confirm the current status of the registrar's accreditation under the ANSI-RAB program.

ABS Quality Evaluations, Inc.
16855 Northchase Drive
Houston, TX 77060
Tel: 713-873-9400
Fax: 713-874-9564

**American Association for
Laboratory Accreditation
(A2LA)**
656 Quince Orchard Road #704
Gaithersburg, MD 20878
Tel: 301-670-1377
Fax: 301-869-1495

British Standards Institute, Inc.
8000 Towers Crescent Drive, Suite 1350
Vienna, VA 22181
Tel: 703-760-7828
Fax: 703-761-2770

**Bureau Veritas Quality
International (NA) Inc.**
509 North Main Street
Jamestown, NY 14701
Tel: 716-484-9002
Fax: 716-484-9003

**Canadian General Standards
Board**
Quality Certification Branch
222 Queen Street, Suite 1402
Ottawa, Ontario, Canada K1A 1G6
Tel: 613-941-8669
Fax: 613-941-8706

**Det norske Veritas (DnV)
Industry, Inc.**
16340 Park Ten Place, Suite 100
Houston, TX 77084
Tel: 713-579-9003
Fax: 713-579-1360

KPMG Quality Registrar
Three Chestnut Ridge Road
Montvale, NJ 07643
Tel: 201-307-7900
Fax: 201-307-7991

**Lloyd's Register Quality
Assurance, Ltd.**
33–41 Newark Street
Hoboken, NJ 07030
Tel: 201-963-1111
Fax: 201-963-3299

National Quality Assurance, U.S.A.
1146 Massachusetts Avenue
Boxborough, MA 01719
Tel: 508-635-9256
Fax: 508-266-1073

National Standards Authority of Ireland (NSAI)
5 Medallion Centre
Greeley Street
Merrimack, NH 03054
Tel: 603-424-7070
Fax: 603-429-1427

Quality Management Institute
Suite 800 Mississauga Executive Centre
Two Robert Speck Parkway
Mississauga, Ontario, Canada
L4Z 1H8
Tel: 416-272-3920
Fax: 416-272-3942

Quality Systems Registrars, Inc.
13873 Park Center Road, Suite 217
Herndon, VA 22071
Tel: 703-478-0241
Fax: 703-478-0645

SGS International Certification Services
1415 Park Avenue
Hoboken, NJ 07030
Tel: 201-792-2400
Fax: 201-792-2558

TUV America, Inc.
5 Cherry Hill Drive
Danvers, MA 01923
Tel: 508-777-7999
Fax: 508-777-8441

TUV Rheinland of North America, Inc.
12 Commerce Road
Newtown, CT 06470
Tel: 203-426-0888
Fax: 203-426-3156

Underwriters Laboratories, Inc.
1285 Walt Whitman Road
Melville, NY 11747
Tel: 516-271-6200
Fax: 516-271-8259

Where to Go
for More Information

ISO 9000 Standards

The ISO 9000 standards, as well as copies of the U.S. counterparts to ISO 9000 known as ASQC Q9000 through Q9004, are available from the American National Standards Institute (ANSI) or the American Society for Quality Control (ASQC).

American Society for Quality Control

The American Society for Quality Control (ASQC) provides information on a range of quality system, quality assurance, and quality control activities, including the ISO 9000 standards. The ASQC can be reached at 611 East Wisconsin Avenue, P.O. Box 3005, Milwaukee, WI 53201. Tel: (800) 248-1946 or (414) 272-8575. Fax: (414) 272-1734. Web: http://www.asqc.org. In addition, there are local ASQC chapters that meet monthly to discuss all aspects of quality. Ask for the local contact and attend one of their meetings to discover the local resources available.

Registrar Accreditation Board

The Registrar Accreditation Board (RAB) evaluates and accredits registrars to perform ISO 9000 audits in the United States, as well as provides information about those accredited registrars. The RAB can be reached at 611 East Wisconsin Avenue, Milwaukee, WI USA 53202. Tel: (414) 272-8575. Fax: (414) 765-8661. Web: http://www.rabnet.com.

American National Standards Institute

The American National Standards Institute (ANSI) administers and coordinates the United States' private sector voluntary standardization system. It was a founding member of the ISO and plays an active role in its governance. ANSI promotes the use of U.S. standards internationally, advocates U.S. policy and technical positions in international and regional standards organizations, and encourages the adoption of international standards as national standards. ANSI can be reached at 11 West 42nd Street, New York, NY 10036. Tel: (212) 642-4900. Fax: (212) 398-0023. Web: http://www.ansi.org.

National Institute of Standards and Technology (NIST)

National Institute of Standards and Technology (NIST) provides information on standards weights and measures useful for calibrating test measurement and inspection equipment to meet the requirements of the ISO 9000 standards. They can be reached on the Web at http://www.nist.gov.

International Organization for Standardization (ISO)

International Organization for Standardization (ISO) is a worldwide federation of national standards bodies from some 100 countries, one member from each country. Its mission is to promote the development of standardization and related activities throughout the world with a view to facilitating the international exchange of goods and services, and to developing cooperation in the spheres of intellectual, scientific, technological, and economic activity. ISO's work results in international agreements which are published as International Standards. ISO can be reached at Case postale 56, CH-1211, Geneva 20, Switzerland. Tel: +41 22 749 01 11. Fax: +41 22 733 34 30. Web: http://www.iso.ch.

CEEM Information Services

Another excellent source of information is CEEM Information Services, 10521 Braddock Road, Fairfax, VA 22032. Tel: (703) 250-5900. Fax: (703) 250-4117. They offer newsletters, manuals, guides, directories, and reports about ISO 9000.

Bibliography

"ISO 9001: Quality systems—Model for quality assurance in design, development, production, installation and servicing," International Organization for Standardization, Geneva, Switzerland, 1994.

"ISO 9002: Quality systems—Model for quality assurance in production and installation," International Organization for Standardization, Geneva, Switzerland, 1994.

"ISO 9003: Quality systems—Model for quality assurance in final inspection and test," International Organization for Standardization, Geneva, Switzerland, 1994.

"ISO 8402: Quality management and quality assurance—Vocabulary," International Organization for Standardization, Geneva, Switzerland, 1992.

"ISO 9000-1: Quality management and quality assurance standards—Part 1: Guidelines for selection and use," International Organization for Standardization, Geneva, Switzerland, 1987.

"ISO 9000-3: Quality management and quality assurance standards—Part 3: Guidelines for the application of ISO 9000 to the development, supply and maintenance of software," International Organization for Standardization, Geneva, Switzerland, 1991.

"ISO 9004-2: Quality management and quality system elements—Part 2: Guidelines for services," International Organization for Standardization, Geneva, Switzerland, 1991.

"ISO 9004-3: Quality management and quality system elements—Part 3: Guidelines for processed materials," International Organization for Standardization, Geneva, Switzerland, 1993.

"ISO 10011-1: Guidelines for auditing quality systems—Part 1: Auditing," International Organization for Standardization, Geneva, Switzerland, 1990.

"ISO 10011-2: Guidelines for auditing quality systems—Part 2: Qualification criteria for auditors," International Organization for Standardization, Geneva, Switzerland, 1991.

"ISO 10011-3: Guidelines for auditing quality systems—Part 3: Managing audit programs," International Organization for Standardization, Geneva, Switzerland, 1991.

"ISO 10012-1: Quality assurance requirements for measuring equipment—Part 1: Management of measuring equipment," International Organization for Standardization, Geneva, Switzerland, 1992.

U.S. CFR (Code of Federal Regulations), Title 21 CFR Parts 210 and 211, Current Good Manufacturing Practice: Amendment of Certain Requirements for Finished Pharmaceuticals, Proposed Rule, U.S. Food and Drug Administration, Docket No. 95N-0362.

U.S. CFR (Code of Federal Regulations), Title 21 CFR Part 820, Good Manufacturing Practice for Medical Devices: General, 1978, U.S. Food and Drug Administration, pages 131–140.

U.S. CFR (Code of Federal Regulations), Title 40 CFR Part 792, Good Laboratory Practices, U.S. Environmental Protection Agency, Office of Information Resources Management.

2185—Good Automated Laboratory Practices, Principles and Guidance to Regulations for Ensuring Data Integrity in Automated Laboratory Operations with Implementation Guidance, U.S. Environmental Protection Agency, Office of Information Resources Management, August 1995,

"Working Draft of the Current Good Manufacturing Practice (CGMP) Final Rule," Docket No. 9ON-0172, Rockville, MD, Office of Compliance, Center for Device and Radiological Health, FDA, 1995.

"Guideline on General Principles of Process Validation," Center for Drugs and Biologics and Center for Devices and Radiological Health Food and Drug Administration, May 1987.

"Everything you always wanted to know about the medical device amendments...and weren't afraid to ask," U.S. Department of Health and Human Services, Public Health Service, Food and Drug Administration, Center for Devices and Radiological Health, HHS Publication FDA 84-4173, second edition, March 1984.

"FDA's New GMP Working Draft: Industry's Last Chance for Comment," by W. Fred Hooten, *Medical Device & Diagnostic Industry,* September 1995, pages 72–79.

"FDA's New Approach to Conducting Device GMP Inspections," by Z. Frank Twardochleb, *Medical Device & Diagnostic Industry,* September 1995, pages 58–66.

"Quality Assurance of Computer Systems: What Is Needed to Comply with ISO 9000, GMP, GLP, and GCP?," by S. H. Segalstad, *Laboratory Automation and Information Management,* 31, 1995, pages 11–24.

"Quality Assurance of Computer Systems Compliance with GMP, GLP, GCP and ISO Standards," by S. H. Segalstad, *LIMS/Letter,* Vol. II, Issue I, Spring 1996, pages 6–7.

"Vendor Audits for Computer Systems: An ISO 9000-3 approach," by S. H. Segalstad, *Laboratory Automation and Information Management,* 32, 1996, pages 23–31.

"Supplier Auditing and Software," by S. H. Segalstad, *European Pharmaceutical Review,* Vol. 1, Issue 3, 1996, pages 37–44.

"A Practical Guide to Validating LIMS," by S. H. Segalstad and Synnevaag, *Laboratory Automation and Information Management,* 26, 1994, pages 1–12.

Index

ABOUT THE AUTHOR

Helen Davys Gillespie is a freelance writer and industry analyst who specializes in high technology. She started her first business in 1991, and now operates Write Away Communications while publishing the LIMS/*Letter* and managing the LIMSource Web site.

For twelve years she worked in the corporate environment as a marketing communications manager, sales support manager, and senior writer/editor for high-technology companies such as Varian Associates, Allergan Humphrey, PacTel Spectrum Systems, and Tymnet. Through her employment with those companies, she gained extensive experience in international marketing and strategic product positioning.

In addition to being a contributing editor for *Today's Chemist* and *Scientific Computing & Automation,* her pieces have appeared in several trade publications such as *PI Quality, Quality Progress, LC/GC, Pharmaceutical Technology, R&D, Environmental Lab, Genetic Engineering News, Inside Laboratory Management, American Laboratory,* and *International Laboratory.* Her articles range from business profiles to features about spectroscopy, chromatography, and laboratory information management systems (LIMS), as well as quality system management issues and regulatory and validation requirements.

In June 1995, she launched a quarterly newsletter for LIMS professionals entitled the LIMS/*Letter.* In June 1996, the LIMS/*Letter* was placed online as a key part of the LIMSource Web site (http://www.LIMSource.com), which comprises some 600 pages of LIMS news and events, including book reviews, product and service listings, application articles, and more.

Her clients include Cirrus Logic, Clontech Laboratories, Digital Equipment, DisCopyLabs, Hewlett-Packard, Lam Research, Nalorac, NovaLink Technologies, Paradigm Technologies, Perkin-Elmer Nelson, Symantec, Thermo Separation Products, and Varian Associates. Her high-technology experience ranges from analytical chemistry to data communications to software and semiconductors. She has assisted Varian Sample Preparation Products, Varian Customer Support, DisCopyLabs, Clontech Laboratories, Cirrus Logic, Paradigm, and Symantec with documenting their ISO 9000 quality manuals.

Ms. Gillespie received a B.A. degree with Honors in English from California State University, Chico, California, and pursued graduate studies in folklore at the University of Sheffield in England. She is a member of the American Society for Quality Control (ASQC), the Commonwealth Club, and the Business Marketing Association (BMA). She has achieved the BMA's Certified Business Communicator (CBC) designation, and has been trained as a Lead Assessor for ISO 9000 quality systems.